Instructional planning systems

A gaming-simulation approach to urban problems

Instructional planning systems

A gaming-simulation approach to urban problems

John L. Taylor

Senior Lecturer in Town and Regional Planning, University of Sheffield and Deputy Director of the Sheffield Centre for Environmental Research.

Cambridge

at the University Press 1971

CAMBRIDGE UNIVERSITY PRESS
Cambridge, New York, Melbourne, Madrid, Cape Town, Singapore, São Paulo, Delhi

Cambridge University Press
The Edinburgh Building, Cambridge CB2 8RU, UK

Published in the United States of America by Cambridge University Press, New York

www.cambridge.org
Information on this title: www.cambridge.org/9780521112734

First published 1971
This digitally printed version 2009

A catalogue record for this publication is available from the British Library

Library of Congress Catalogue Card Number: 70–160093

ISBN 978-0-521-08229-7 hardback
ISBN 978-0-521-11273-4 paperback

Contents

Figures

Tables

Forms

Acknowledgements

This book has been made possible through the co-operation and generous assistance of a number of agencies and many individuals, and to all these supporters the writer would like to express his sincere thanks and gratitude.

Much of the stimulus for this work was derived from the author's studies in the United States, and the influence of American scholars and practitioners has permeated and shaped not only this research but the whole instructional simulation field. To these pioneers a great debt is owed and the writer must count himself fortunate in being able to draw upon a great deal of this collective effort through the channels of personal friendship. In this context, special personal thanks are due to all fellow research workers in the field and, in particular, to R. H. R. Armstrong and Margaret Hobson of the University of Birmingham, K. R. Carter of Manchester Polytechnic, Professors R. D. Duke and F. Goodman of the University of Michigan, Professor A. G. Feldt of Cornell University, Dr R. N. Maddison, Dr N. Rackham, both formerly of the University of Sheffield and last, but by no means least, to Rex Walford of Maria Grey College, London.

To the Royal Institute of British Architects and the Department of Town and Regional Planning, University of Sheffield, much more than the traditional thanks for their financial support is owed. Their assistance has been both timely and generous, especially bearing in mind the novelty and innovative character of this work. Without their backing and encouragement it is difficult to conceive how such an effort could properly proceed. The continuous co-operation available within the University of Sheffield deserves particular mention. The University granted permission to draw upon material from the writer's Ph.D. dissertation ' Some Aspects of an Instructional Simulation Approach to the Urban Development Process '. Professor J. R. James, head of the Department of Town and Regional Planning and colleagues within this department have remained tolerant mentors, indulgent supporters and provocative critics. Special recognition is extended to the individuals concerned as well as to the University's external examiners Professor Gerald Dix and Professor Peter Haggett who as a group, in various ways, stimulated the writer to extend and set down, in book form, the fruits of on-going exploration in the realm of urban development gaming-simulation.

At the same time a number of other agencies have freely given of their time and expertise in supporting the writer's exploratory studies. These agencies include too many professional institutes, research centres, universities and local authorities to be identified individually but special thanks must be extended to associates in the Organization for Economic Co-operation and Development, the Town and Country Planning Summer School and the Salzburg Congress in Urban Planning and Development.

Appreciation is also expressed to numerous long-suffering colleagues and students who not only encouraged the writer in his investigations but also acted as guinea pigs in various trials and experiments throughout the last five years. This work is a testimony to their enthusiasm and patience and it is hoped that the end product is worthy of their help and commitment.

Particular acknowledgement is made to my wife and family for the tolerance and good grace which is always required in supporting efforts of this nature. Here the load has fallen on my wife who, throughout this project, has acted as willing experimental subject, secretarial assistant and reader.

Finally, and by no means least, a considerable debt is owed to the secretarial and technical staff of the Department of Town and Regional Planning. Their hard work has improved the presentation of this document and has greatly expedited its completion.

To all the above and to the many others who have freely contributed in some way the writer would wish to renew his thanks and sincere appreciation.

University of Sheffield JOHN L. TAYLOR
December 1970

1. Point of departure

This work aims to record the development and relevance of a new simulation approach to the study of the urban development process at university level. Before examining gaming-simulation, the instructional technique which is the core of this approach, an attempt is made to put this new aspect of educational technology into some overall perspective. First, this is done with reference to the dynamics of the general educational situation and second, it is attempted with regard to the evolving state of education for planning. Then the role of gaming-simulation models, as a major element in certain instructional simulation systems, is tentatively outlined. Finally, after posing some of the questions which represent the starting point of this study the intentions and methods of approach are summarized.

THE EDUCATIONAL CONTEXT

How can university education be related to a rapidly changing world? What can the new educational technologies offer in meeting today's problems and tomorrow's needs? Where can the performance of higher education be improved?

This type of question, and many others are, quite rightly, often being presented by educators in search of change and improved performance. The concern here, however, is somewhat limited in that much must be left unsaid in order to take a closer look at the attitudes and climate which surround innovations and experiments in higher education. Obviously, there are many attitudes which inhibit the use and development of innovatory instructional systems at this level and some of these issues are now briefly considered.

The role of the university teacher tends to be a rather limited one; promotion and status derive from scholarship, research or practice rather than from teaching skill (Hale Committee 1964); the traditional emphasis is on 'reading for a degree'; the ritualistically enshrined timetable divisions and the reliance upon sessional examinations tend to obstruct change and experimentation; inter-university exchange and collaboration on a formalized basis is rare; and the time and resources available for the exploration of new teaching methods and the development of new curriculum units is not generally available. In addition, training opportunities for university teachers are limited and there are distinct interdisciplinary problems involved in the subject specialist's attempts to enlist the aid of the psychologists and specialists in education. These, then, are a few of the problems which confront the would-be pedagogic innovator.

On a more hopeful note there are several encouraging signs that the above position is changing and some of these issues are now presented in an effort to arrive at a more balanced overview. First, it is important to note that enquiry or research into teaching in higher education is rapidly gaining momentum (Beard 1970). At the same time interest in teaching methods at this level is expanding and larger numbers of teachers are now taking part in teaching enquiries or experiments. Second, and leading on from the first point, the ' liberal ideas movement ' in university education does appear to be gaining some ground. For some this has meant: a search for new means of tackling academic problems; a re-examination of traditional methods and, for others, an analysis of emerging aspects of educational technology. In short, new methods, new materials, new subjects and hence even new objectives and goals are being more commonly defined and being given a wider airing.

As outgrowths from this ' movement ' some of the new British universities have established staff teacher training and service centres. Some of these units have equal status with other departments and are intended to provide courses and information for university colleagues as well as acting as a focus for carrying out, co-ordinating or communicating research and development in education. It is on this point that there is obviously much room for improvement. So far the university teacher has not appeared to greet innovation and experiment with either the zest or enthusiasm which perhaps the topic deserves. New methods and the liberal education movement are not enough in themselves unless they have an impact on everyday routines and procedures. The shortcoming is often that new methods and the boldness of a few are not sufficiently applied or communicated to the wider audience. There is a need for a greater awareness of new techniques and new thinking. In sum, there is a need for wider recognition and deeper understanding of educational frontiers and research work in progress.

Following from this point it must be emphasized that a total or blind commitment to innovation and experimentation is not being advocated. The interest and questioning which is attached to new approaches to learning should of course, be also directed to time-honoured ways. It is not that all is known about these ways or that they are questionable but that, especially in times of rapid change, these methods should be under continuous review. Thus, the present research is founded not on the belief that the present systems are bad, although this may be the case, but rather that there is a natural desire to search for improvement and better systems.

EDUCATION FOR PLANNING

Attitudes which inhibit change in the higher education system are applicable equally to planning education and perhaps additional drawbacks are suffered by virtue of the discipline's comparatively short history. The profession, as yet, is a small and very young calling with strictly limited resources at its disposal. Within such a fledgling profession more attention appears to have been focused upon increasing the numbers being trained than on the content of courses and the methods of teaching. Recently, however, a move to clarify institutional objectives at a national level has been made and certain syllabus components have been defined (T.P.I. 1969). It now remains to be seen whether the means to achieve these re-evaluated ' ends ' can be fashioned with equal academic and professional backing.

The unprecedented expansion of education for planning which began in the 1960s has brought both problems and opportunities. Staff recruitment appears to have followed rather than preceded the expansion in student numbers and a shortage of experienced or well-qualified teachers is everywhere apparent. The relatively sudden increase in the number of institutions offering planning courses cannot, therefore, be unreservedly welcomed. It is accepted that there is a desperate need for planners and hence a pressing need for more training opportunities. However, the proliferation of new centres may in the short-term do little to improve the overall quality of planning instruction. Many of the established institutes would not claim to be well-equipped or staffed and the advent of several competitors can only mean an increasing fragmentation of already scarce resources with fewer opportunities for innovation and experiment backed by proper financial assistance, competent staffing and full supporting facilities.

Accompanying this upsurge of interest in planning has come a new educational picture with reference to urban and regional studies. Traditionally, town and country planners have been recruited from the ranks of graduates in civil engineering, architecture, surveying and the social sciences; and given a full or part-time course to gain professional qualification. Latterly, with the expansion of education for planning, a large number of new full-time undergraduate and postgraduate courses in urban and regional planning have been established. In parallel with this, the physical planning component of many architectural, engineering and surveying courses has been strengthened. Further, the field of physical planning is one which has found a place in the curricula of the social sciences, particularly in economics, sociology and geography courses. One of the most important results of this widening awareness of urban and regional planning has been that increasing interdisciplinary expertise has been brought to bear on environmental problems. New skills and new levels of understanding have been derived from multiple fields and a narrow definition of planning is no longer possible. In particular, increasing agreement has been reached in differentiating in the planning field between conceptualizing urban and regional systems on the one hand, and the control and guidance of the evolution of these systems on the other. The impact of the ' systems ' approach to planning has been tremendous (Chadwick 1966, and McLoughlin 1967 and 1969). In short, it has freed planners from the mechanistic, deterministic view of the city and region which saw these entities as artefacts, composed of bricks and mortar, drains, roads and water pipes. Now cities and regions are seen increasingly as a set of complex interactions between social and economic dynamics and the physical artefacts. The planner's job of making alterations to the disposition of physical elements in the city or region has gradually become one of defining the political, behavioural, social, economic and technological generators of systems evolution and adapting the physical structure accordingly; realizing, of course, that changing the physical elements will have repercussions on the rest of the system.

Therefore, to define the objectives of a planner's education at a high level of generality, one could say that it is to enable students to discover a systemic overview of the city and region and to acquire the understandings and skills necessary to manage the process of developmental change. It is in this context that simulation will be considered.

Thus simulation may be said to have a two-fold relevance to planning corresponding to the twin objectives defined above. First, simulation is perhaps the only way in which a systemic view of a complex city or region can be economically conveyed. This explains partly why an increasing number of simulation models of urban and regional systems have been developed recently (Mitchell 1961, and Harris ed. 1965). Few of these, however, have been designed for *instructional* purposes, for the most part they are research tools. This discussion, however, concentrates on the simulation models which have been developed for instructional purposes. Second, simulation presents opportunities for learning how to guide and manage the human settlement. Via this medium the student is free to interact with the environment of the simulation model; so that he discovers both the dimensions of the model and the extent of his influence on model variables. Thus most of the simulations to be discussed below were designed with these objectives in mind and are a direct response to the need to improve the quality of current understanding and practice.

WHY GAMING-SIMULATION?

In recent years greater stress has been placed on the need to consider ways of structuring planning teaching so that clearly defined objectives can be achieved. The best of traditional studio situations are experimental: the student can study in varying depths whatever aspect of the situation engages his attention. There are occasions, however, when one might want to ensure that the student perceives the situation more systematically. One way of describing the situation and giving it a particular content is through the medium of gaming-simulation. Not all the elements of an urban system can be expressed readily in mathematical terms, with, for example, the optimum situation expressed in a linear programme. The involvement of people, as individuals and socially through institutional, administrative and political processes means that a human, non-rational element is a crucial part of the system. This is particularly the case when one thinks of planning as guiding and managing the evolution of the urban (or regional) system. Gaming-simulation, with its mixed strategy approach to the coverage of the quantitative, through mathematical terms, and the qualitative, through human representation, appears to be an appropriate vehicle for conveying such an understanding.

In short the real world is frequently regarded as a far from ideal training ground. As it is expensive, difficult and sometimes dangerous to study or experiment with reality, it is natural that some researchers should have sought alternative and more amenable ways of learning. What the potential planner needs, amongst other things, is a learning environment or laboratory where he is free to experiment and come to terms with the operational complexities of reality or hypothetical ' futures '. A dynamic environment is required which does not interrupt or interfere with reality and yet is more explicit and performs in a comparable manner to the real or hypothetical system. With gaming-simulation procedures in mind Meier (1963, p. 350) has elucidated on this point:

> The real challenge is to reproduce the essential features of a city in a tiny comprehensive package. A set of maps is not enough. Years must be compressed into hours or even minutes, the number of actors must be reduced to the handful that can be accommodated in a laboratory or classroom, the physical structure must be reproduced on a table top, the historical background and law must be synopsized so that it can become familiar within days or weeks, and the interaction must remain simple enough so that it can be comprehended by a single brain. This last feature is the most difficult challenge of all!

Put another way, what are constantly being sought by teachers in professional planning programmes are better methods for conveying a wealth of diverse material in a coherent form. Such methods have to communicate experience as well as facts and have to reconcile conflicting as well as overlapping theory. They must be capable of abstracting and representing, dynamically, the essentials of any situation so that they describe and respond in accord with the milieu within which the planner works.

THE APPROACH

Before going on to examine the problem in greater depth the writer's viewpoint must be made clear. First, the fundamental reason for studying a gaming-simulation approach to the urban development process lies in a personal belief that the technique has some potential, that it is worthy of development, and that in certain circumstances particular models of this type can favourably affect student learning. The material presented here is an attempt to look objectively at instructional systems whose base, from the outset, was thought to have some value; this book is essentially an examination of this conviction.

Second, from the initiation of the writer's investigations, early in 1966, it seemed clear that the technique was never likely to occupy more than a minor place in the planning studies timetable – something which was a possible means of supplementing rather than supplanting other forms of instruction. The result is an analysis of an important, yet modest, avenue of instruction which can perhaps have a useful auxiliary role in urban development studies.

Third, following on from the preceding point, it must be realized that the discussion concentrates upon only *one* branch of simulation and more particularly is concerned, very largely, with instructional aspects of urban development gaming within the higher education system. This is not to suggest that other applications of these techniques are not of equal importance or that some of the information presented here is not relevant to related fields of study. The particular focus of attention, in the writer's opinion, is an underdeveloped branch of instructional simulation which is viewed neither as an exclusive path to greater student understanding nor as a technique without problems of its own.

In addition to these basic assumptions and predispositions with regard to gaming-simulation technique, a series of premises concerning the individual's conception of education for planning should be explicitly, but briefly described. The writer tends towards the view that there is no *one* way of training potential planners nor is there any single type of planner. In other words, there is room in the planning educational system for a diversity of approach and for a variety of planning expertise. Some students need and desire to be taught, even within one institution, in perhaps many different ways. This need has to be recognized even though it may not always be possible for it to be satisfied in whole or in part.

As an extension of the preceding point it is also believed that a number of diverse skills can advance our understanding of the urban scene and no one discipline appears to have more than a partial answer to many urban problems. Here, however, a subject specialist view is presented as a contribution to the continuing dialogue amongst the many individuals and groups involved in both the urban development and the educational process. The writer hopes to have benefited from many authorities, fellow research workers, and students in these two areas and certainly has enjoyed the stimulus from the growing numbers endeavouring to fashion more effective instruments of learning. Obviously publications of this nature are only made possible through this type of support and co-operation and formal acknowledgement has already been gratefully made to the particular contributions of certain individuals and institutions. Despite this support, any errors and shortcomings to be found here are regrettably the author's responsibility, although it is hoped that the extensive source material identified, at the end of this work, will serve to reduce major omissions.

This publication, then, is concerned with the evolution of urban and regional instructional simulation systems and their relevance to planning and the study of urbanism. In addition to the confidence vested in the potential of this form of approach there is by implication a further source of motivation which must be stated. The rapid upsurge of interest in gaming techniques and the proliferation of planning models has left little time for practitioners or academics to take stock of the instructional simulation process in terms of its growing theory, methodology, or operational formulations. Similarly, it is felt that there has been a noticeable lack of philosophical reflection and semantic clarification regarding the international development of planning games and related simulation procedures. Thus, mindful of the fact that as the number and variety of gaming-simulation models being used and developed increases, so the problem of establishing a balanced overview becomes more essential and at the same time more difficult, this account is restricted to an examination of the first decade in the history of planning games. As previously much of the literature has dwelt on specific gaming models and their application in experimental situations, here a somewhat wider view of the ' state of the art ' is attempted.

In particular, it is thought that so far little attention has been directed towards avoidance of certain popular misconceptions and establishment of wider understanding. To date, no comprehensive picture of activities and knowledge concerning gaming-simulation of the urban development process has been available other than that scattered throughout various journals, institutional monographs and personal correspondence. Thus, there has been little clarification of the extent to which one new methodological approach to land use planning instruction has been developed, discarded or utilized on an international scale and at various academic levels. Other fundamental problems can be raised in the form of such questions as: Why use instructional simulation? What, when and where should gaming procedures be used? And how do planning games function and inter-relate with other activities? Such unknowns are the starting point of this study and in attempting to supply answers to the questions posed, it is hoped that some light will be shed on a new aspect of educational technology relevant to the study of the urban development process. The work will have served some purpose if it points the way for further research and more balanced discussion and, hopefully, wider experimentation in education for planning.

2. The basis for discussion

This chapter sets out to: define the area of study; clarify the relevant terminology; and identify the book's major levels of concern. In particular it attempts to provide answers to the questions: What is a gaming-simulation model? What should be in mind when the term planning game is encountered? And what is meant by an instructional simulation system?

PLANNING AND THE URBAN DEVELOPMENT PROCESS

Planning is often used as a generic term to cover certain parts of the decision-making process but in this work, unless otherwise specified, it is used to denote those types of planning concerned with land use allocation and the shaping of the built environment. More particularly, comments are very largely restricted to what in Britain is more popularly known as 'town and country planning' and in the United States as 'city and regional planning'. Here for clarity and brevity the terms 'urban planning', or simply 'planning' are usually used to encompass both terms as well as to cover the compromise title of urban and regional planning.

Urban planning involves many skills and diversely qualified personnel. To avoid narrow definitions the confines of limited professional and institutional orientations are, as far as possible, subjugated in favour of a more comprehensive approach to the urban development process. The book seeks to avoid demarcation disputes of the 'who should do what' variety in an attempt to take an overall view of the management of the urban process. Thus, it is problems rather than professional terms of reference which are the focus of this discussion.

SIMULATION

Any attempt at simulation definitions invariably leads to semantic difficulties. Thus the definitions and descriptions offered here have attempted to bear this in mind and endeavour to avoid the dogmatic and contentious statement.

The bewildering confusion of terms and definitions, and the lack of a commonly accepted vocabulary results partly because so many diverse disciplinary antecedents are involved, partly from the complexity of the issues under consideration and partly also from the increasing amount of independent and often isolated research being undertaken. Individuals as well as institutions have felt free to draw upon separate semantic sources to suit their convenience and local as well as disciplinary meanings have been established as and when the need arose. So it is that war gaming, social science, computer technology, operational research and game theory have all contributed towards an embarrassingly rich technical vocabulary.

Obviously, it would be presumptuous to try and settle once and for all the terminology involved in a relatively young and rapidly developing field. At this point, however, it is necessary to make certain distinctions clear and inevitably the definitions used here may not *always* coincide with those presented elsewhere. The deciding factors in establishing certain meanings have been clarity, simplicity and the apparent weight of current usage.

The use of the term ' simulation ' as an all-embracing word is becoming very popular and, as it means different things to different people, is perhaps already overworked. In a general sense the word delineates a range of dynamic representations, or models, that employ substitute elements to replace real world or hypothetical components. It should be noted, straightaway, that throughout this book the phrase ' real world ' is used as a generality to refer to that which is being simulated. As Loewenstein (1966, p. 112) has pointed out, every model in some sense ' simulates ' reality by way of an abstraction. Whatever form of substitution is involved, the ultimate objective is the construction of an easily manipulated system in order to facilitate study.

Hartman (1966, p. 4) has simply defined ' simulation ' as: ' the development and use of models for the study of the dynamics of existing or hypothesised systems '. To be useful this definition must be expanded and qualified according to certain specified criteria. A workable elaboration is provided by the System Development Corporation (1965, p. 2):

> the systematic abstraction and partial duplication of a phenomenon, activity or operation to effect:
>
> 1. the design of a system in terms of certain conditions, behaviour and mechanisms,
> 2. the analysis of a specific phenomenon, or
> 3. the transfer of training from a synthetic environment to a real environment.

Three separate purposes are indicated in the above simulation definition: design, analysis and training. In practice all simulations do not always fall neatly into these divisions and a combination of objectives is possible. To some extent the author's interests, in currently investigating the educational uses of simulation techniques, encompass all three purposes.

A further definition of simulation and a good guide to some of its properties has been provided by Monroe (1968, pp. 6–7):

> a simulation is small in comparison to the real system; it can be begun and stopped at will; it is readily observable.

The properties of a simulation are of three types:

1. iconic properties, transformed in scale but otherwise the same as the properties they represent;
2. analogue properties, in which one property is substituted for another but behaves in the same way as the original property; and
3. homologue properties in which one property is substituted for another, but the substitute bears only a surface similarity to the property it represents.

In addition to the classification by objective, purpose or property it is also possible to categorize simulation techniques in terms of other criteria. For example, System Development Corporation's System Simulation Research Laboratory has classified its simulations according to the relative degree of involvement of people and equipment (Redgrave 1962, p. 6). Alternatively I. J. Good (1954, pp. 68–9) and John Moss (1958, p. 591) have classified simulations according to their degree of abstraction from the real life system, operation or procedure. Such breakdowns are merely illustrative of ways in which simulation studies might be ordered and a more complete taxonomy of simulation types can be found elsewhere (Harman 1961).

The specific form of simulation about to be discussed is generally referred to as gaming-simulation. To put this activity into some overall perspective and to clarify the relationship of this branch of simulation to pedagogic methods it is necessary to return to some of the qualifications of simulation already touched upon in the preceding discussion. If, for explanatory purposes, the last mentioned simulation classification is adopted, as in Figure 1, then a single scale of related instructional simulation techniques is established with respect to their degrees of abstraction and the following major elements are encountered:

(a) *Case studies*

These exercises involve detailed descriptions or histories of selected problem situations which the student is required to analyse and discuss. The approach was very largely pioneered by the Harvard Law School and was subsequently developed and popularized by the staff and students of the Harvard Graduate School of Business Administration (MacNair 1954). As Forbes (1965, p. 15) has pointed out, no matter how realistic the symbolism might be surrounding decision issues in a case study it is essentially a static device, for once questions of choice have been made and discussed the case model has served its usefulness.

(b) *'In-Basket' or 'In-Tray' methods*

Basically this is a method for studying the responses and the decision-making ability of individuals in specific situations. 'In-basket' situations extend case

Figure 1. *Some related simulation techniques.*

<REALITY — INCREASING ABSTRACTION>

CASE STUDY	<–> IN-BASKET OR IN-TRAY METHOD	<–> INCIDENT PROCESS	<–> ROLE PLAYING	<–> GAMING SIMULATION OR GAME SIMULATION	<–> MACHINE OR COMPUTER SIMULATION
OBSERVATIONS ON THE REAL WORLD	NON INTERACTING ONE TO ONE REPRESENTATION	INTERACTING ONE TO ONE REPRESENTATION	INFORMALLY STRUCTURED GROUP PORTRAYAL	STRUCTURED GROUP REPRESENTATION	ALL DATA & DECISIONS EMBEDDED IN A MATHEMATICAL REPRESENTATION

After Duke & Burkhalter (1966, p. 2).

studies to a more practical level; the individual is called upon to play one specific role and must, in isolation, act on a number of hypothetical issues raised by the morning post. The participant is given a set time to make decisions arising out of each item of correspondence and these responses provide a basis for performance assessment. The technique was pioneered and has been highly developed by the Educational Testing Service, Princeton, New Jersey, as a test for use in selecting American Air Force Officers (Zoll 1966, pp. 6–11). The main feature of the test is the presentation of realistic problems in such a way as to elicit what might be termed reasonable responses. In other words, the examinee must perform just as he would on the job – writing letters or memoranda and issuing guidance and directions, just as if he were at his own desk. Consequently, interest not only rests on what is done, but also on the manner of execution (Reed 1966, p. 16).

(c) *'Incident' processes*

This technique is a modification of the case study method and it has been developed and well described by Paul and Faith Pigors (1961) of M.I.T. In the incident process, the student is not given the entire case material and thus, throughout the exercise, must seek additional information by question and answer. In this way, the incident approach adds a data collection task to the case study process of analysis and discussion. Here, however, greater emphasis is placed on the student's assessment of what is required to formulate a decision and being able not only to recognize this need, but also to call precisely for this data requirement.

(d) *Role playing*

A process requiring spontaneous mock performances from a group of participants in which an attempt is made to create realistic and life-like situations (Cole 1961, p. 34). The people involved are required to act out problems of human relations and then must discuss the development of their interaction, frequently involving conflict, with other role players and observers. It is a technique which is used to gain empathy with prescribed roles as well as to gain insight into human interplay in the context of a safe learning environment and relies on spontaneous enactments to illustrate and dramatize human problems or actions. It should be noted that role playing, socio-drama and psychodrama are terms for closely related techniques which are sometimes confused but well differentiated by Moreno (1947) and Garvey (1967).

(e) *Gaming-simulation or game-simulation*

Rauner and Steger (1961, p. 3) have indicated that RAND adopted the latter expression to describe studies which incorporate both the free, exploratory, relatively unstructured characteristics of business or war-games and the rigid, controlled, well-structured qualities of traditional computer simulations. Such

RAND studies customarily consist of human decision-makers interacting with a simulated environment; this environment is represented by other humans in combination with various models of the real world. The participants are then confronted with varied situations in order to facilitate the study of human behaviour or the development of particular processes. These gaming procedures will be elaborated in later sections of this chapter. However, at this point, it should be noted that in addition to the term 'gaming-simulation' two of the more common synonyms for the technique are educational gaming or operational gaming. As some authors (Thomas and Deemer 1957 and Cohen *et al.* 1964, for example) reserve the term 'operational gaming' for those models designed to attempt to find an optimal solution to a game this particular synonym will not be used here. Instead, the terms gaming-simulation and educational gaming will be used interchangeably as generic names for the family of procedures of which the planning game is but one specific branch.

(f) *Machine or computer simulation*

This species of simulation, as the name suggests, is used to identify operational models that have been programmed for high speed computing equipment. Here the model is operated by manipulating the various symbols and programmes which replicate the variables and components of the system. Consequently, in these simulations, all data and decisions are embedded in the machine and considerable precision is achieved when handling quantifiable factors (Dawson 1962, pp. 8–9 and Brody 1963, p. 197).

Before trying to establish a deeper understanding of gaming-simulation, in particular, it should be emphasized that the preceding treatment of related instructional simulation techniques is in no way comprehensive. There is, of course, considerable overlap amongst the constituent procedures and the continuum presented here does not claim to be a completely adequate taxonomy of simulation types but hopefully serves as a simple ordering device to clarify the ensuing discussion.

It should also be noted, at this point, that the foregoing range of instructional simulation techniques are not commonly found, either singly or in various combinations, in many urban studies programmes. With regard to in-basket, incident and role-playing approaches, it is not without significance that the author is not aware of any literature specially covering the application of these techniques to land use planning instruction. Certainly they are used to a limited extent in dealing with, for example, development control problems from the single, straightforward application for planning permission through to mock appeals or public enquiries involving major policy decisions (see for example, Table 1). However, these different yet related simulation techniques must be left without further comment in order to turn

Table 1. Outline of a Maria Grey College role-playing exercise

Milford Haven Planning Authority

(Constituted 1968)

Chairman :
Alderman Miss D. WEDGEBURY
Member of Pembroke C.C.

Members of Committee :

Miss KANJI	Rep. Milford Fish Federation.
Mrs. V. HIGLETT ...	Rep. S. Wales Oil Co's Fed.
Miss P. RULE	Chairman, Milford U.D.C.
Miss L. GOLDRIDGE ...	Chairman, Pembroke U.D.C.
Miss S. HERMES ...	Chairman, Neyland U.D.C.
Mrs. L. BLACKWELL ...	Chairman, Pembroke R.D.C.
Miss C. WATKINS ...	Hon. Sec. Pembrokeshire Coast National Parks Committee.
Miss A. GRIFFITHS ...	Warden, Dale Fort Field Centre (also member of H. West R.D.C.)
Miss S. TURNER ...	Rep. Ministry of Development.
Miss A. ELLIS	Rep. H.M. Welsh Office.

The next meeting of the above will be held on Monday, June 24th 1968 at 9.30 a.m. at Maria Grey College, at which your presence is requested.

AGENDA

1. Welcome to Delegates.
2. Resumé by chairman of development of Haven 1800–1968.
3. Proposals;
   (1) Application by Esso to extend Herbrandston Refinery to double size at Sandy Haven.
   (2) Application by Pembroke R.D.C. to authorize Holiday Camp for 3,000 at Angle Bay.
4. Statements by U.D.C. Chairmen on the needs of their districts.
5. Comments on item 4, by representatives of H.M. Government.
6. Any other business.

R. Walford, Hon. Secretary. 17.6.68

Source : Walford (1968)

to the gaming-simulation procedures which constitute the fundamental structure of planning games.

GAMING-SIMULATION

A clearer appreciation of gaming-simulation procedures and planning games can perhaps best be obtained by: first expanding upon the earlier outline description of gaming-simulation; second, to avoid any confusion, clarifying what game theory involves; third, summarizing the sequence of events which often constitute an academic game; fourth, and finally, describing some of the core features commonly found in planning games.

Gaming-simulation has already been described, briefly, as one part of a spectrum of related simulation techniques. At this point a further clarification of the terminology is important not only because of the technique's various synonyms: academic gaming, operational gaming, and the less common, heuristic gaming; but also because of the widespread confusion and seemingly liberal use of the words 'game', 'gaming-simulation' and 'simulation'. The games which are the concern of this investigation are structured exercises with a simulated environment. In other words, the game resources, constraints and goals have been largely determined in relation to prescribed real-world or defined hypothetical systems. Thus a simulation procedure or learning sequence is built up in such a way that it resembles the form of a game.

For example, in the writer's LAND USE GAMING-SIMULATION, (LUGS), described in Appendix 1, just as in its parent model the COMMUNITY LAND USE GAME (CLUG), players acting as entrepreneurs buy, sell and develop land according to the modified 'laws of the market place'. Here the modifications represent normal statutes and conventional town and country planning controls. The game model structures bidding and implementation procedures without unduly limiting the opportunities for competition, co-operation, collusion and conflict. There are, just as in every gaming-simulation, definite relationships between player decisions and game results; sometimes these relationships are made explicit and sometimes they must be deduced. In the course of acquiring a particular expertise or group of skills and attitudes incremental learning is encouraged by continuous feedback on the adequacy of performance throughout the educational process. Decisions generally have a bearing on the state of the simulated environment and participant results are influenced not only by his own decisions but often by those of other players.

In brief, games of the above type are a form of simulation which involve the participation of human actors, generally in a competitive situation. Gaming-simulation differs from other forms of simulation largely because of its reliance on human decision-makers as integral parts of the simulated system and because of its relatively low level of precision. Thus gaming-simulations are didactic instruments involving a concise and cumulative presentation of a situation which might be too dynamic, too disordered and too complex to be represented economically by other means. As may be recalled from the spectrum of related simulation techniques set out in Figure 1, a distinction was made between gaming-simulation and a number of other forms of instructional simulation. Throughout this discussion these distinctions are relied upon, and the point to note here is that all gaming in this context is a form of simulation, but not *all* simulations are gaming-simulations.

Two comments regarding the wider use of the word ' game ' now seem to be in order. First, many writers have been reluctant to accept the choice of the term for a pedagogic activity when quite naturally the word has a strong connotation with levity and entertainment, in the main, connections hardly likely to elicit respect and support from the uninitiated (Greenlaw *et al.* 1962, p. 5, and Pugh 1965, p. 1). As a consequence, to avoid any confusion with parlour games or sporting pursuits, and to be more in line with a serious educational approach, the word is often abandoned in favour of more attractive qualifications for the term ' simulation '. Obviously learning can be enjoyable as well as serious and however unfortunate the label it does apparently have some value as it has been seen to be attractive to some students who might be diffident or repelled by a more formidable title (Monroe 1968, p. 17).

Second, a clear distinction should be made between planning games and all gaming-simulations, on the one hand, and ' the theory of games ' or ' game theory ' on the other. The latter two terms have their roots in the work of John Von Neumann who from 1927 onwards was concerned with the logic of conflict and the evolution of a theory of strategy. This work culminated in the publishing of the classic text *Theory of Games and Economic Behaviour* (Von Neumann and Morgenstern 1944) and led, amongst other things, to renewed interest in conflict analysis and strategic planning. However, at the moment, the theory of games has little direct applicability for gaming-simulation users although it has undoubtedly stimulated serious interest in games *per se*.

In drawing a distinction between ' game theory ' and ' gaming-simulation ' the former might better be termed ' mathematical game theory ' and although this work will *not* be considering the theory it seems necessary, in avoiding any confusion, to elaborate on its meaning. Mathematical game theory is concerned with the formal analysis of a great variety of conflict or potential conflict situations. It is essentially an analytical approach to decision-making and the resultant consequences of a range or combination of actions. Game theory seeks to discover the logical structure of conflict situations and describe them in mathematical terms. As Press and Adrian (1966, p. 5) have reminded us, rarely can it prescribe a definite strategy although, through implication, it can perhaps clarify some of the critical conditions under which choice should be made clear.

There are numerous books on game theory to suit all levels. Williams (1954) has provided but one of the many popularizations of Von Neumann and Morgenstern's classic text; Luce and Raiffa (1957) have comprehensively surveyed the field of games and decisions; Rapoport (1960 and 1965) has pursued a continuing interest in what he terms ' fights, games and debates '; and recently Peston and Coddington (1967) have produced a basic ' primer ' as an introduction to the essentials of game theory. The burden of all these works and the related literature concerns conflict and co-operative behaviour and attempts to *prescribe* the choice or combination of choices which might lead to the best ' pay-off ' in set circumstances. This means that much of ' game theory ' research is concerned with the *definition* of concepts such as coalition formation, strategy determination and conflict values.

While at the moment game theory offers very little at an operational level for planning games it is evident that it may offer more in the future. This confidence stems from the work of an increasing number of geographers and regional scientists who have made direct use of Von Neumann and Morgenstern's ideas in a study of land use patterns. Doreen Massey (1968) has recently summarized some of these efforts in relation to problems of location and has highlighted the promising initiatives of Stevens (1961), Gould (1963), Isard (1967) and Isard and Smith (1967). Despite the limited immediate practical value that can be found here, the theoretical potential seems worthy of further investigation, and already work in this direction can be credited with having done much to widen the conception of the word ' game ' beyond its more popular connotations of a frivolous kind.

After this slight, but necessary, digression it is now time to return to the nature of gaming-simulation procedures as well as their characteristics and functions. A clearer appreciation of these procedures can perhaps best be obtained by outlining the sequence of events which often constitute an academic game. In short, prospective participants (players) are familiarized with the details of the actual exercise (the game). Certain objectives are usually described and actual play begins with an ordered system of decision-making. Decision outcomes (pay-offs) are almost immediately fed back to the players for evaluation. The cycle of decision-making, feedback and evaluation is repeated to allow the equivalent of many years of decisions to be completed in a single day. During this time, administrators closely observe all proceedings to enable further feedback on the players' actions and performance to be presented at the termination of the game. At this stage, a post-mortem or critique session is held to discuss the gaming process and the development of particular strategies. The players have an opportunity for reviewing their performance and discussing the recurrent effect of continuous feedback. Finally, actual results and human interactions are analysed by the administrators to clarify and reinforce lessons learnt during play.

The simulation games used by urban planners and those in related disciplines vary widely in their details. However, in common with many other applications of instructional simulation systems they often appear to share the following core features:

(a) They are principally used for learning how systems react under continually changing conditions.

(b) They are in essence simple abstractions of relatively complex aspects of hypothetical or real world situations.

(c) They very largely achieve their simplicity through reducing complex operations into a series of simply expressed actions controlled by explicit rules.

(d) They expose participants to certain pre-selected features under relatively controlled and risk-free circumstances.

(e) They allow the concerted use of physical models, mathematical representations and human operators.

(f) They require participants to: assume roles involving various degrees of co-operation, competition and conflict between players or teams; and make decisions which reflect their understanding of key features of the model.

(g) They produce certain decision 'pay-offs' – rewards or deprivations – determined by chance, by reference to human assessments or alternatively by the use of predetermined rules and formulae.

(h) They provide varied experience in controlling the course of events over a series of 'time' spans where the state of the simulated environment is continuously altered in response to the quality of accumulated decision-making.

(i) They generally compress 'time' and, as a result, are able to provide rapid feedback on the results and consequences of decisions.

(j) They progress in predetermined decision stages or periods and each period generally represents an allotted 'time' span during which multiple, and sometimes simultaneous interaction between co-operating and competing decision agents takes place.

The essence of gaming-simulation, in the context of this investigation, is that it represents an experimental attempt to develop an instructional tool to approach the decision-making activities and consequences involved in the urban development process. Individuals are involved in 'engineered' situations where a number of outcomes are possible and although they separately or jointly exert some control they do not have complete control over each other. The game merely simplifies the complexities of the situation and allows easy comprehension and manipulation of otherwise intricate mechanisms. In these circumstances gaming procedures provide a framework for organizing and rationalizing a diversity of data and non-concurrent theories concerned with urban development. The form of the framework can be orientated towards the inherent structure of the phenomena or those structures which attempt to rationalize the process of understanding. To date, the majority of planning games have adopted the latter approach although such efforts as the METRO programme (Duke 1966) are making a very real attempt to represent a specific sub-region and to simulate the growth patterns which *might* actually occur. However, with reference to this investigation the main body of interest rests with less sophisticated instruments for transmitting some of the broader concepts of urban development at a university level of interest.

More specifically, although gaming simulation techniques lend themselves to an assortment of instructional uses, three areas of higher educational usage are already discernible in the fields of teaching, training and research. So far, for planners, the teaching and training functions have attracted the greatest interest although most models developed to date lend themselves, with varying degrees of suitability, to a multiplicity of uses in all three areas.

In summary, then, simulation games are gross operational replicas that endeavour to provide insights into the dynamics of an ongoing system. The participants are provided with decision-making experience over an extended period of simulated time, within a controlled and risk-free environment. The game replaces the complexity of the real world with a simplified abstraction which allows certain representative features to be easily understood and readily manipulated. There is quick feedback on all decisions and the games are structured into periods which force participants to repeatedly consider the development cycle of a series of inter-related problems. The technique provides a synthetic experience which allows players to profit from being directly exposed to, and personally involved with, preselected features of a rapidly evolving system. Basically, gaming-simulations are abstractions or simplifications which in this context are designed to advance appropriate learning in relation to the urban development process.

Having completed a general introductory account of gaming-simulation techniques it must be acknowledged that such statements and even detailed descriptions of particular operational models are not substitutes for first hand involvement. For this reason the diligent reader yet to encounter instructional simulation procedures in action *or* lacking experience with the application of the technique to urban problems is advised to turn to Appendix 1 before proceeding further. This Appendix provides introductory guidance for organizing an elementary simulation exercise designed to demonstrate basic operational gaming procedures to those unfamiliar with the technique in a planning context.

INSTRUCTIONAL SIMULATION SYSTEMS

Finally, a note of explanation is called for in relation to the umbrella phrase 'instructional simulation system'. Here this phrase is used to embrace a comprehensive package of learning materials and techniques; a 'mix' of resources which may vary from one situation to another but above all is a complete pedagogic system built up of integrated components whether they be particular processes, media or other aids. At least two points emerge from such a conception. First, the operational game is but one part of the gaming-simulation process and cannot be viewed in isolation, nor can it stand alone. Second, and following on from this point, any instructional simulation approach must be related to a curriculum context and the total 'mix' of educational resources.

To elaborate on the first point, the instructional simulation systems in mind throughout this discussion can be treated as falling naturally into four stages. The first stage covers the model building and overall design process, the second stage covers the briefing and background preparation essential to the third stage, which involves the actual operation of the model, and finally comes the de-briefing and general evaluation stage. These stages are overlapping and inter-woven into a total experience and in the better situations it should be difficult to distinguish between each stage as a separate entity. Similarly, in the writer's opinion, the whole instructional simulation system embodying the gaming process should ideally be integrated into the curriculum so that it is not distinctly recognizable as a separate teaching unit.

Thus, using a gaming model as an instructional instrument involves a sequence of learning experiences where the actual game is but one component part of a total effort to maximize the use of educational manpower, plant and techniques. For example: at the briefing stage, a gaming-simulation exercise may encompass varying forms of programmed learning supplemented by such teaching aids as film loops and audio tapes. The operational simulation may be a testing ground for introducing a battery of decision aids and planning techniques such as 'Discount Cash Flow Accounting', 'Planning Programming Budgeting Systems' and 'Critical Path Studies'; while the actual mechanics of the exercise may confront the participant with a range of 'interfacial' technologies centred on such things as: data banks, computerized accounting procedures and computer graphics systems. Finally, at the de-briefing and evaluation stage closed circuit television and the video tape may be some of the visual analytical aids which may usefully be employed in capitalizing upon feedback sessions.

From the foregoing outline illustration it will be clear that a gaming-simulation model is *not* seen as an educational tool in itself just as an instructional simulation system is *not* thought to be a complete training programme.

This publication examines an approach to the urban process which, although dependent upon gaming-simulation procedures, also relies upon the carefully planned utilization of the total armoury of educational resources. Therefore instructional simulations are considered here in a ' systems ' context where the manner of use and the wider relationships are just as important as the central component – which, in this case, happens to be a gaming-simulation model.

So, after having attempted to establish certain working definitions and an understanding of specific pedagogic procedures, it is now time to turn to an examination of the general historical background to the evolution of certain instructional simulation systems.

3. The genesis of an approach

This section traces the genesis of instructional simulation systems with specific reference to the development of planning games. A direct linear progression of events which precede and underpin the advent of this form of simulation is discernible but their inter-relationship and the effect of some concurrent activities is highly debatable. More specifically the roots of the urban development gaming-simulation are examined with reference to three streams of endeavour involving: military innovators, social scientists, and a small, yet significant, interdisciplinary group of academic theorists.

WAR GAMING

Gaming-simulation procedures have a venerable pedigree in the form of war games. But, despite their age and popularity as well as the fact that these military games might be regarded as the earliest known form of instructional simulation, there are few broad surveys of the field or treatises on their interrelationship with the rapidly developing area of social science simulation.

Perhaps the most recent and certainly one of the most readable accounts of the origins and development of war gaming has been provided by Wilson (1968) in his book *The Bomb and the Computer*. Despite its informal and circumscribed title this work is a succinct and comprehensive account of the military commitment to games. Other useful overviews of the field have been presented during the last twenty years by, amongst others: Young (1956, 1957 and 1959), Thomas (1957) and Cohen and Rhenman (1961). Articles by these authors have included historical surveys of war gaming but have not attempted to provide a contemporary assessment of ' the state of the art '. In the apparent absence of this literature Quade (1964) has assembled a broad cross-section of manuscripts from practitioners in the field and from contributions by authorities such as Abt (1964), Shephard (1970), Specht (1957) and Weiner (1959) it is possible to assemble an outline picture of some aspects of work in progress.

When, where and how military gaming originated is unclear. Many authors have pointed out the obvious similarity and common historical background between war games, chess and related board games but the history of such developments is not well documented. In Murray's (1952) definitive work on the history of board games he was careful to recognize that if a link existed, it was not one of levity but of common concern and interest in conflict situations. Certainly there is a great deal in common between chess and many of the older training games used as symbolic equivalents of warfare (Murray 1913, pp. 46–50). From beginnings of this nature it was, as Cohen and Rhenman (1961, p. 132) have pointed out, probably a very short step to attempt to use the medium for planning and training purposes.

In particular, Wilson (1968, p. 1) credits two sources with being possible forebears of war gaming contests. One source is the Chinese game of ' Wei-Hai ' (meaning ' encirclement ') which originated about 3000 B.C. and has its modern equivalent in the Japanese game of ' Go '. It was, and still is, played on a stylized base board with different coloured stones and the objective is simply to outflank your opponent. The other source is a Hindu battle game called ' Chaturanga ' which was played by four persons on a hypothetical map with military pieces and the outcome of moves was determined by the throw of a dice.

Of greater interest here, however, is the branch of the military game movement (Neues Kriegsspiel) which resulted from the growing belief that war was rapidly becoming a science and therefore its conduct was subject to the application of scientific principles. By the turn of the eighteenth century this movement had taken hold of the Prussian militia and highly formalized and elaborate procedures were evolved to encompass a wide range of tactics and logistics. The continuing military search for even greater accuracy and realism led to complications and a split in the gaming movement resulted. After 1870, formal distinctions between ' free ' and ' rigid ' Kriegsspiel were clearly discernible. The former approach tended to rely upon the judgements of experienced umpires while the latter relied heavily upon detailed rules and formal procedures.

From this period onwards interest in the technique noticeably widened and an international clientele was quickly built up. Two world wars heightened the level of interest and the emergence of Operational Research coupled with the advent of computers led to developments of increasing complexity and sophistication. From the broadening scope of these achievements the political-military game can be seen as but one of the natural outgrowths. These derivatives are about to be discussed but before leaving the subject of war games it is important to stress at least three points which should have emerged from this short survey.

First, military leaders have long recognized the instructional ability of war games, and military training through gaming has probably been going on for many centuries and for at least three centuries has been extensively developed and used (Cohen and Rhenman 1961 and McKenney 1967). Second, over time the medium has shown its value and versatility in keeping pace with changing needs, so much so that interest in the utility of the technique continues to flourish. For example, in 1963 according to Wilson (1968, p. 47) there were listed officially some 200 gaming models, excluding many used purely for military training at lower levels. At the same time the same author went on to state his belief that this number was probably being doubled at about two-yearly intervals! Third, and finally, a substantial investment in terms of time, money and energy has been expended in developing war games, and contemporary commitments are no less modest as the seven million dollar U.S. Navy Electronic Warfare Simulation (NEWS) bears witness (Greene and Sisson 1959, p. 1, and Kibbee *et al.* 1961, p. 6).

SOCIAL SCIENCE SIMULATION

The second influential area in the development of planning games has been the social sciences. With the exception of Guetzkow's (1962) 'Simulation in Social Science' and Raser's (1969) 'Simulation and Society' general summaries of this collective effort are unfortunately rare. However, there is little doubt that the major contributions have come from gaming-simulation interests in business, politics and education. This research is now outlined together with what might best be termed the 'fall-out' from certain separate but closely related initiatives.

Business games are the oldest form of social science game and by far the greatest body of gaming-simulation literature is concerned with these applications. In Robinson's (1966, p. 89) opinion there are probably as many business games as there are games in all other educational areas put together. Certainly the literature dealing with business and management gaming is extensive although, as far as the writer can detect, no single agency or author has attempted to document this wealth of information on a continuing basis. In its place there are spasmodic and unco-ordinated 'state of the art' accounts to be found in conference proceedings and the periodic reports of educational innovators. At least four symposia deserve mention: the National Symposium on Management Games (1959); the American Management Association Meeting (1961); the Ford Foundation and Tulane University Conference (Dill *et al.* 1961); and the Association of Teachers of Management Report (1965). During this period 'benchmark' publications of equal significance have come from such innovators as: Greene and Sisson (1959); Kibbee *et al.* (1961); Greenlaw *et al.* (1962); and Fairhead *et al.* (1965).

In addition, a rash of post-war simulation progress reports on individual games has emerged accompanied by extensive classified references to many of the business and management models that are in use. Guides to this literature are to be found in works by: Loveluck (undated), Greenlaw *et al.* (1962), Rohn (1964) and Newbold (1964). At a more specific level there is an equal number of well-documented case studies of the design, development and use of particular models, for example: Cohen *et al.* (1964) on the CARNEGIE TECHNICAL MANAGEMENT GAME; Thorelli and Graves (1964) on the INTERNATIONAL OPERATIONS SIMULATION of the University of Chicago; and McKenney (1967) on the use of H.B.S. SIMULATION GAME at the Harvard

Graduate School of Business. Each of these publications provides an excellent introduction to the topic under discussion.

Business games owe their existence largely to the foresight and resources of the American Management Association (A.M.A.) which in 1956 initiated the development of the first widely publicized business game. This is not to suggest that the idea had not been mooted earlier, indeed as Greenwald (1966, p. 17) has pointed out, Norman Angell conceived the idea of a MONEY GAME in 1928 and this probably represents one of the first references to the instructional simulation of a monetary system. However, it was left to the A.M.A. to recognize and develop the link between the war game and big business. Thus in 1956, after an exploratory visit to the U.S. Naval War College, a research group was formed, by the A.M.A., to design the first management game as a direct outgrowth of military gaming experience (Ricciardi *et al.* 1957).

From this point on, mainly as a result of the very enthusiastic reception accorded to A.M.A.'s TOP MANAGEMENT DECISION SIMULATION, business gaming became a widespread training activity for universities as well as for industry and commerce. In just over a decade a variety of games has been developed which range from fairly simple decision exercises lasting little more than an hour through to extremely complex simulations involving anything up to several days to cycle and in case of the CARNEGIE TECHNICAL GAME occupying over 50% of the second year of their MBA course (Greenwald 1966, p. 19). Among other things, these games aim to communicate management principles and business skills in such diverse areas as: marketing, production, stock control and labour relations. Although the present number of games is already very large, and shows every sign of being on the increase, it is possible to generalize on some aspects of their utility and characteristics. For example, over one hundred educational, industrial and government organizations are now using gaming as an experimental, teaching, or operational device (Loveluck undated, Dale and Klasson 1964, and Shubik 1968). Despite the lack of validation studies the consensus has been that in all these areas the technique has been of some value. The resources involved probably rival the extent of the military investment in the technique and certainly represent a testimony to the training fervour of the post-war business world.

Aside from the A.M.A.'s leadership and the buoyancy of modern business it merits a slight digression, at this stage, to identify certain other factors which have not only accelerated the development of business games but also have influenced the whole evolution of instructional simulation systems. Both endeavours have benefited from, amongst other things, recent advances in computer technology, the continued development of a theory of games and a widening interest in operational research. The impact of these different, yet closely related, factors is in few ways concrete and yet they have obviously influenced both the philosophy and technology surrounding the wider application of quantitative methods. From the employment of these techniques a ' systems approach ' to problems has emerged which among other things has found some value in gaming-simulation as a tool to increase human understanding of complex situations. Basically, the concern here is with the build-up and application of a scientific method and whilst these developments in themselves are not perhaps of outstanding significance they have collectively stimulated great interest and at the same time have encouraged further research – particularly in the fields under discussion.

The ' spill-over ' from the development of these techniques, outside business, has been felt most noticeably in the realm of political gaming. Here games for both teaching and research have proliferated quickly and have been diffused over a wide area of interests. A large part of this effort is a direct outgrowth from war gaming and owes much to the influence of ' think-tank ' specialists cultivated chiefly by the RAND Corporation. RAND's interest may be viewed as a natural extension of their extensive involvement with sophisticated military games and a furtherance of their previous, elaborate experimentation with gaming-simulation techniques. In 1954 their institutional gaming was so successful, as a procedure for the study of foreign affairs, that the technique was soon adopted and developed elsewhere (Goldhammer and Speier 1959). Several American simulation research programmes were set up and especially noteworthy developments were carried out by Harold Guetzkow (1963) and his colleagues at North-Western University, and Lincoln Bloomfield (1959, 1960 and 1965) and associates at Massachusetts Institute of Technology. More recently, European experimentation and development has been undertaken by Banks, Groom and Oppenheim (1968 and 1970) at University College, London. Although all these continuing research efforts have been concerned with the simulation of international political situations these centres are not the only exponents of the technique nor are all political games carried out at this level. For example, Burgess (1966) has identified other political-diplomatic games such as THE CRISIS GAME: SIMULATING INTERNATIONAL CONFLICT (Griffin 1965) and hybrid conflict games concentrating on sensitive problem areas such as Vietnam (Laulicht and Martin 1966). In addition there are simulations of federal government (Garvey 1965), state legislatures (Boocock and Schild 1968) and local political systems (Goodman 1968, and Davison 1961). Certainly in the United States instructional political systems are rapidly becoming almost as numerous and popular as gaming-simulation in the business sector. Here the considerable backing of American ' think-tank ' and educational technology industries are evident, although references to the level of this investment are lacking.

Aside from the influential and widespread diffusion of games for business and political purposes other social scientists have not been slow to experiment with the technique. However, in comparison with these two fields other disciplinary applications are less extensive if no less successful. For example: Paterson (1970) has built up considerable experience with industrial relations games; Forbes (1963) has designed THE COLLEGE AND UNIVERSITY PLANNING GAME as a simple mechanism for classifying some of the values underlying academic growth; THE HOSPITAL EXERCISE IN LONG TERM PLANNING (P.E. Consulting Group 1966) is a training game currently being used by training boards and extramural departments in the United Kingdom; Kitchen (1970) has described British European Airways' experience with gaming-simulations for the training of control staff; and finally Rae (1969) has outlined a series of gaming exercises as part of a liberal studies programme for fine art students. This is by no means an exhaustive listing and certainly does little justice to a growing number of gaming models employed by Colleges of Education as most recently described by McLeish (1970). However, it should be emphasized that the main objective in presenting this and the material in the preceding sections is not to produce a definitive survey of simulation usage but to indicate some of the many ways the technique is being used as well as identifying major sources of further information. Wherever possible, references have been selected to show that urban development issues have commanded some simulation attention even before a distinct variety of planning game emerged.

EDUCATIONAL INNOVATIONS

Accompanying the recent evolution of military and social science gaming has been a series of parallel simulation developments in the general field of education. The better known of these innovations have involved role playing, psycho-drama, in-baskets, incident processes or T-groups – techniques concerned with the transmission of performance skills, the creation and expression of basic attitudes as opposed to the complete concentration on the communication or gathering of factual information. In spite of their many differences, all of the above forms of educational simulation have served to extend the interest in a related range of instructional simulations which, among other things, have added strength to gaming-simulation.

Much of the credit for these initiatives must go to social psychologists, experimental psychologists and sociologists who from the 1930s onwards were developing face to face confrontations as laboratories for the study of group behaviour. In this context simulation and gaming were readily accepted as vehicles for extending research into human behaviour in varied learning environments. One product of this interest was the extension of diplomatic and military simulation exercises from higher education to the school system. Other games followed this lead as teachers began to succumb to the enthusiasm of university behavioural scientists or began to search for new orientations to learning in reaction to the weight of formal, passive or rote learning traditions.

The resultant development of school games with simulated environments has been headed by such centres as: Johns Hopkins, where under the direction of James Coleman (1961, 1966 and 1968) several gaming models are in various stages of design, development and production; Northwestern University, where Harold Guetzkow (1962, 1963 and 1964) created the INTER-NATION SIMULATION and subsequent school derivatives; Abt Associates, in Cambridge, Massachusetts, where a variety of elementary, junior and senior high school, college and university games have been designed, primarily on a contractual basis, under the leadership of its founder Clark Abt (1966); and the Western Behavioral Sciences Institute at La Jolla, California, where Project Simile is directed by Hall Sprague (1966) and is concerned with an exploration of the educational uses of simulation in social studies, primarily at secondary school level. These centres and others like them, together with countless individual schools and teachers, have quickly built up a variety of gaming models for use over a wide range of age and ability levels; indeed, American school system games probably outnumber and certainly have a longer history than planning games (Boocock and Schild 1968, and Tansey and Unwin 1969). The growth of interest in academic games, particularly in secondary schools, has obviously been dramatic and in the United States appears to have almost reached the stage where school simulations underpin much of the instructional simulation developments at university level (Zieler 1969).

Of particular interest to this discussion is the growing popularity of academic games in the urban studies sector of the school system. For example: Abt Associates have designed POLLUTION and NEIGHBORHOOD as community simulations for the Wellesley, Massachusetts, Public School System (Kaplan 1966*a*). Similarly the game of MANCHESTER (Blaxall 1965) was designed by Abt under contract to Educational Services Incorporated and the game of SECTION was designed by the same company for the High School Geography Project at Clark University (Kaplan 1966*b*). The High School Curriculum Development Project of the Association of American Geographers have also built up a 'Growth of Cities Unit' which incorporates the PORTSVILLE game (A.A.G. 1967). Although most of the pioneering work on this type of procedure, at a school level of interest, has been carried out in the United States, some work on geographical games has been done in Britain. For example, Smith and Cole (1967), two geographers at Nottingham University, formulated a set of notes on gaming ideas to test out teachers' reactions and their bulletins stimulated sufficient interest to encourage the production of a primary school series 'New Ways in Geography' which relies heavily upon gaming material (Cole and Beynon 1969). A similar consumer probe was carried out by Rex Walford (1968) of Maria Grey College, London, who issued a comparable set of notes on 'Six Classroom Games for Use in Geography Teaching' and such was the response that these games and the ideas behind them are now explained in a succinct handbook entitled *Games in Geography* (Walford 1969). To date, the impact of this last group of innovators has been slight. However, from these modest beginnings it is not too much to hope that, over time, a powerful force for change may spread throughout the entire education system.

So far, then, this chapter has been about the roots of planning games in both military and social science simulation. In addition, certain educational innovations have been identified as sources from which all gaming-simulation has benefited and from which urban and regional instructional simulation systems have drawn or are likely to draw particular support. After these surveys it is time to turn to an examination of the background to planning games in terms of the theoretical contributions from a small and yet significant group of scholars who appear to have done much to speed the advent and development of simulation gaming in planning; some of these contributions are now considered.

THEORETICAL EXPLORATION

As early as 1958, Meier, a central figure in the evolution of planning games, began to provide some of the groundwork for the development of new and wider applications of simulation systems and in four articles in *Behavioral Science* (1958*a*, 1958*b*, 1959 and 1961) he discussed, with considerable foresight, the emerging form and content of some aspects of social science theory. The last article in this series (Meier 1961, pp. 232–48), sought an explanation for the then concentration on management and military simulation when the application of the technique to other areas was just as relevant. In particular, Meier considered the educational and training potential of gaming-simulation in understanding basic social, political and cultural institutions. He suggested several possible development areas which included the simulation of: ecological evolution and population growth; community administration and planning; and public service operation. Meier was careful to point out that the foregoing suggestions, taken from a more extensive list, represented only a limited sample of directions for future research. However, subsequent events were to fall within, or follow, these remarks so it is now proposed to examine, in turn, some of the progress which helped to advance or consolidate work in these areas.

Meier and his associates at the School of Natural Resources and Mental Health Research Institute, University of Michigan, were responsible for devoting considerable attention to the translation of theory into practice within their own fields. Their concern for ecology and the training of conservation students generated at least three operational gaming models: BIG

SHOOT, WILDLIFE and the WATERSHED MANAGEMENT GAME (Duke 1964, p. 10). In the WILDLIFE series of games (Meier *et al.* 1964, pp. 67–76), various aspects of population dynamics are communicated via simulation exercises as an alternative to presentations in purely mathematical or verbal terms. The resulting games against nature are viewed by the designers as chess board applications of Simon's (1957) theory relating to mechanisms for choice in a risk-laden environment. Their interest, for this discussion, lies in the number of avenues they open up for further development – work which still, very largely, awaits to be done.

The basic WILDLIFE model, developed by Meier and Doyle, embodies the concept of community in its elementary sense (Meier *et al.* 1964, and Meier 1968). A year of natural history can be played out within an hour, through representations of the thoroughly studied 'moose-beaver-wolf-vegetation systems' of the Isle Royal National Park in Lake Superior. The impact of such incidents as forest fires and epidemics can be traced, and new populations in the form of tourists or indigenous ethnic groups can be added. In its present form the game is essentially a computer based exercise and its future utility appears to depend upon rationalization of machine procedures and the man–machine interface.

Community administration and planning simulations were tackled separately, at a theoretical level, by at least two groups of ambitious scholars. Edwards and Francis (undated), after preparing several competitive games ranging in size from individual free choice games to company and industry simulations, propounded an extraordinarily complex and imaginative model which they called the INTER-CITY COMPETITION: THE COMMUNITY GROWTH GAME. The specific form of the game was designed in response to a need, pointed out by Meier, for a game to improve the utility to the community of time presently spent in inter-school and inter-collegiate sports. The resultant elaborate conception is an attempt to formulate a multi-dimensional contest of linked games representing facets of change and growth in urban settlements over the working life span of the average citizen. Viewed, by the designers, as a vehicle for sociological experimentation and teaching it is unfortunate that, as yet, there appears to be no evidence of efforts to implement these bold ideas.

Similarly, Grundstein and Kehl (1959) have described equally exciting conceptual outlines for elaborate community simulations which also await realization. Grundstein's COMMUNITY GAME proposal is centred on a hypothetical city possessing certain well-defined characteristics and the complete game is intended to simulate the administration of a community over a thirty year period. In terms of scale and intricacy the project is extremely ambitious and therefore it is perhaps not surprising that it still awaits implementation. Despite this drawback Grundstein has gone on to develop his theoretical propositions in a number of challenging articles on some of the conceptual problems associated with the simulation of public social systems (Grundstein 1961, 1966 and 1967) and their wide circulation and profound influence can hardly be under-estimated.

Equally stimulating theoretical contributions have also come from at least two other groups of authors in what Meier termed the public service field. Coleman (1961 and 1965) and his colleagues at Johns Hopkins University have drawn attention to the functional possibilities of games in the School System and have established a centre for the development of academic games. With the assistance of funds from the Carnegie Foundation, of New York, they have been actively concerned with the translation of theory into practice and a number of games from the Centre for the Study of Social Organisation of Schools are now in various stages of design, development and production (Coleman *et al.* 1968*a*, 1968*b*, and Boocock and Schild 1968). Similarly, Bruner, with support from the Ford Foundation, has directed the Social Studies Curriculum Program with equal vigour and a positive commitment towards the exploration of the academic utility of games. His concern has been with the general advancement of educational innovation and theory and his efforts have done much to ensure a wider receptiveness towards instructional simulation systems (Bruner 1961 and 1966).

Finally, there remains the large number of individual authors who have examined games and play in attempts to identify fundamental lessons to be drawn from their relationship with various institutions and social interactions. Most recently, Inbar (1969) has provided a useful review of some of these efforts with regard to games viewed as a play upon life, in general, and as a basis for the study of social organization, in particular. Eloquent pleas for greater research into this ' neglected ' area have come lately from a ' school ' of sociologists and social psychologists who have viewed games as potential vehicles for the advancement of social theory and as a means of investigating the functioning of systems of rules and types of role (Suits 1967*a*, 1967*b*, Coleman 1968*a*, and Inbar 1969). However, it seems clear that Rossi (1957) and Long (1958) were among the first to take pains to link this type of work with the study of community level decision-making within group contexts and with the planning process by mapping out an ecology of local community games as a guide to achieving a better understanding of community development. Whilst Rossi and Long seem to have received some acknowledgement from the pioneer planning gamers (Duke and Feldt personal communication) it is obviously invidious to single out any one authority in a brief and selective review covering such a rich proliferation of ideas. However, in an effort to draw together the preceding amalgam of empirical developments and theoretical propositions a number of other qualifications would now seem to be in order.

The factors just considered, in varying degrees, appear to have played some part in creating a base from which planning games began to emerge; however, the genesis of gaming-simulation in urban and regional studies is by no means clear-cut or a straightforward linear progression of one evolutionary development after another. Recent history, and here the concern is with events of the last decade, is difficult to unravel, partly because of the personal involvement and interpretation available to the historian. Objectivity in such circumstances is therefore difficult, yet throughout the development of instructional simulation systems at least one characteristic does appear to be prominent, that is: the diversity of interest which has been focused upon simulation and the willingness of both designers and operators to draw, unhesitatingly and sometimes indiscriminately, upon almost any branch of the literature appearing to fit their need.

No attempt has so far been made to clarify the degree of overlap between what, for ease of presentation, has been termed the three main streams of gaming-simulation development relative to urban affairs teaching. For the moment, at a high degree of generality, a synoptic view of the background to planning games has been attempted. First, the war gaming antecedents were acknowledged as were related outgrowths from these promising efforts; second, the technical contributions of early game users have been outlined alongside the work of associated social scientists; finally an effort has been made to distinguish what efforts were underway, at a theoretical level, to establish some kind of relationship between certain new analytical techniques (involving amongst other things, decision theory, organization theory and systems theory) on the one hand, and planning and the social sciences on the other. In short, one means of drawing many of these elements together was seen to be the simulation game. These remarks, then, are not intended to provide, at this stage, a rationale for such a pedagogic activity but, rather to make the reader aware of certain features of a ' skeleton framework ' supporting and surrounding some of the pioneer planning games now about to be considered.

4. Selected planning developments

Instructional simulation systems currently being evolved to study the urban process have been seen to have their origins in a number of streams of intellectual endeavour. Drawing freely from these precedents, during the last decade, the planning profession and related disciplines have sought to develop gaming-simulation procedures as instructional instruments for a variety of reasons and functions. It is the historical documentation of such investigations that is the concern of this chapter. But before turning to this task, it must be explained that a comparable chronicle of planning game development is not known to the writer and what overview material is available is first identified.

GENERAL OVERVIEW MATERIAL

There is a paucity of literature attempting to trace the development of planning and related urban affairs games which Taylor (1969), Twelker (1969) and Werner and Werner (1969) have shown is by no means typical of the application of gaming-simulation techniques to other areas. This investigation has already identified extensive source material concerned with military, social science and educational gaming-simulations and has drawn attention to the wealth of business simulation literature. Thus, part of the rationale behind this investigation was to establish an overall view of the growth of a seemingly neglected branch of instructional simulation in an attempt to fill a gap in contemporary planning knowledge.

Previous to the publication of this account, material concerning pioneering planning simulation games had to be pieced together mostly from the work of Duke (1964), Duke and Schmidt (1965), Feldt (1966*a* and 1966*b*), and Meier and Duke (1966). To update these early authoritative reports it is perhaps best to turn to a number of symposium and annual conference reports which, from time to time, have recorded progress in the dissemination of ideas and the build up of operational expertise in this field. Although these events are generally less concerned with establishing historical perspectives and more commonly orientated towards operating or ‘ workshop ’ sessions, in the absence of more systematic assessments, they are important for their pronouncements on planning game development. At a general level, the gatherings of the American Society of Planning Officials (Duke 1965, and Meier 1965), the American Institute of Planners (Patterson 1968) and the Town and Country Planning Summer School (Carter and Taylor 1968, and Taylor 1968) are useful reference points in this respect. More specific and extensive material is to be found in the proceedings of more narrowly conceived meetings, such as P.T.R.C.'s (1967) ' Planning Games Seminar ' held at the University of Manchester or the more recent joint University of Birmingham and Sheffield symposia on ' Instructional Simulation Systems in Higher Education ' (Taylor and Carter 1969*a* and 1971). Obviously this does not identify the sum total of all references in the field but hopefully it serves to indicate the need for more concise and definitive coverage of the development of planning games. Accordingly the rest of this chapter is addressed to a selected presentation of such historical material.

PIONEERING PLANNING GAMES

The adaptation of instructional simulation systems to the study of urban phenomena owes much to planners in the United States. Hendricks (1960) was one of the American pioneers in this field. His demonstrational game, which was the first widely publicized model of its kind, was followed by two other urban development games constructed independently by Duke (1964) and Feldt (1965). Encouraged by enthusiastic reports from these three authors, the planning profession and related disciplines are increasingly undertaking similar research efforts. As a result various games are being used on an international scale by universities, colleges, research organizations and planning agencies.

All three prototype planning games are, in many ways, derivatives of business simulation exercises which in turn are a direct outgrowth from war games. The pathfinder authors freely acknowledge the support derived from these and related endeavours, some of which have already been touched upon in the previous chapter. However, the lead and impetus gained from military as well as social science precedents and associated developments in no way overshadows the importance of the pioneering work carried out by Hendricks, Duke and Feldt. In fact, the approach to the urban process which is the basis of this publication rests very much upon, and has been strongly influenced by, the dynamic set in motion and sustained by these three American planners.

POGE (PLANNING OPERATIONAL GAMING EXPERIMENT) was devised by Hendricks (1960) as a demonstration of gaming techniques for the members of the North California Chapter of the American Institute of Planners. The action of the game is centred upon conflicting interests of staff planners in a mythical planning department, and local property developers. The planners, representing the professional advisers of a small community, are motivated by their desire to ensure good planning and indirectly the continued existence of their office. On the other hand, the land owners are concerned with fruitful speculative dealings and, in particular, with attempts to prevent the adoption of land use regulations contrary to their interests. Both sides are supplied with a list of strategies which they may use in designated areas and specified situations. The third participant is the Planning Commission or the arbitrator whose function is to make known to the players the consequences of their selected strategies. 'Pay-offs', or values assigned to the outcomes of each move of the game, are worked out in detail by the administrator according to a prescribed range of results. Monetary values are attached to specified outcomes and success for both planners and property owners is measured in financial terms. The object of the game is to expose the student to selected situations which involve conflict of interest relevant to the land use planning process.

Figure 2. METROPOLIS *static relationships*

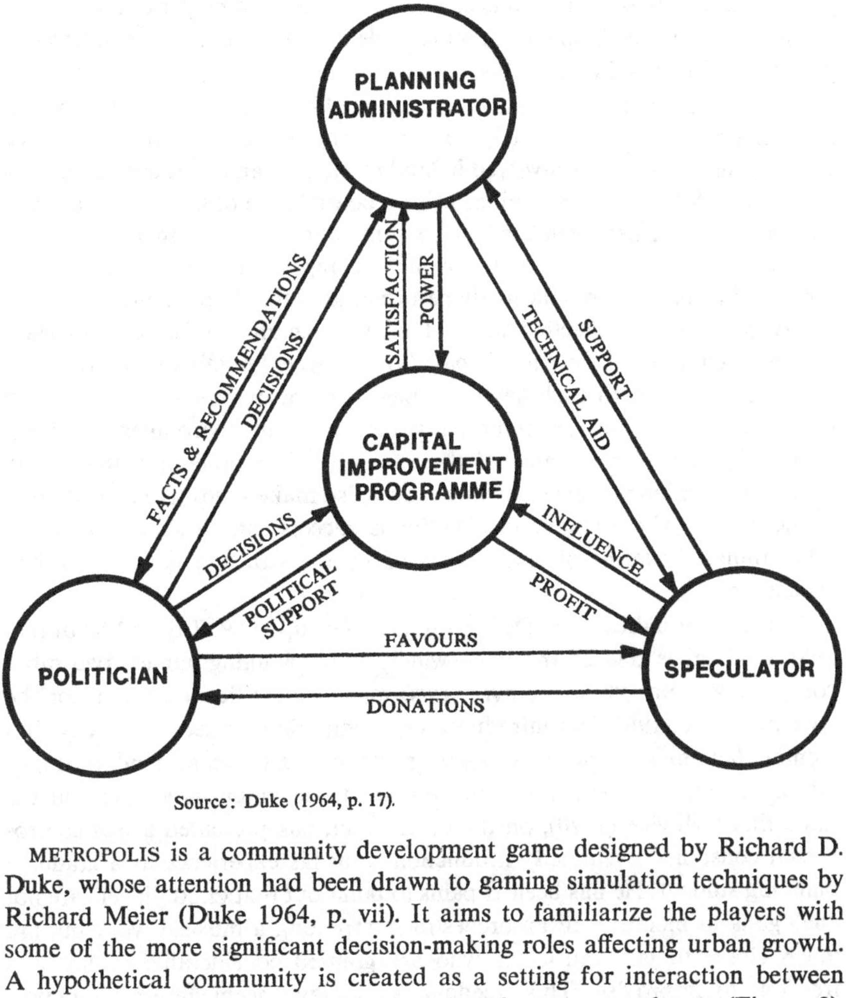

Source: Duke (1964, p. 17).

METROPOLIS is a community development game designed by Richard D. Duke, whose attention had been drawn to gaming simulation techniques by Richard Meier (Duke 1964, p. vii). It aims to familiarize the players with some of the more significant decision-making roles affecting urban growth. A hypothetical community is created as a setting for interaction between planning administrators, politicians and real-estate speculators (Figure 2).

Emphasis is placed on the roles of the players, their relationships with each other and with the form and nature of their environment. Planning administrators gain by accurately projecting revenues and recommending a capital budget which wins legislative approval. Politicians gain by keeping taxes low whilst, at the same time, endeavouring to meet all community needs. Speculators gain by investing money in areas where public works expenditure will particularly enhance land values. Simultaneous decisions are required from each group on two issues in each round. One decision involves a private commitment and is directed toward personal gain. The second decision concerns a public issue and involves community interests. Resources are limited and each move results in direct and accumulative consequences. A number of growth indicators is computed each cycle, as a result of player decisions, and in the final analysis these indicators provide a measure of success in terms of population gains and losses.

CLUG (the CORNELL LAND USE GAME designed by Allan G. Feldt (1965) and now renamed the COMMUNITY LAND USE GAME) provides an analogue of the interactions and changes involved in land economics and land use determination at a high level of abstraction. It is a ' board ' type of game, with specific and fairly rigid rules, which might be compared to a combination of chess and monopoly. The rules of the game are representative of some of the economic forces which tend to shape urban growth. By reducing the multiplicity of variables affecting urban land to a small number of important factors, each move of the game is based on a selected number of alternatives. Play commences with each having a fixed amount of capital which he may use to further personal and community interests in an environmental context defined by the administrator of the exercise. With prudent management, teams can maximize their investments and also make a positive contribution to the growth of the community. Ultimate success can be gauged, amongst other things, in terms of team assets or by the size and nature of urban expansion.

As there are indications (Design Methods Group 1969, Taylor 1969*a*) that Feldt's model is one of the most widely used planning games, two other COMMUNITY LAND USE GAME overviews warrant identification. First, for the layman, Bruce (1967) has introduced CLUG initiation procedures as environmental planning struggles in which the protagonists, or teams, tend to behave like unscrupulous speculators and militant trade unionists at one and the same time! Blivice (1970), on the other hand, has presented a less controversial consumer-orientated introduction from his experience as a graduate planning student. He has been at pains to point out that CLUG players are not analogous to members of an orchestra interpreting a musical work but are much closer to jazz musicians who are granted considerable licence and freedom to improvise. This freedom, all reviews seem agreed, is CLUG's greatest strength. It allows ample opportunity for improvisation without impairing the game's operational simplicity.

All of the three games just described have much in common. They create a model world which works like the real one in particular ways deemed to be appropriate to planning. These miniaturizations can be stopped or started at will and a good deal of the contemporary information overload is withheld. Players operate as businessmen, industrialists, politicians, administrators, householders and planners with varying social and economic motives. What cannot be represented by roles, rules and procedures is handled by the operator who acts as the external world and certain 'Acts of God'. In all three cases play proceeds in a series of steps which make up rounds of rapid decision-making involving numerous cumulative consequences. The teams can choose whether to collaborate or compete with each other and their fortunes are not independent of other decision-makers or the fortunes of the community as a whole. The common challenge is to discover how to manipulate and manage the simulated system before being tempted to tamper with the prevailing order. In short, in each case the player is encouraged to learn what is at stake through progressive stages of experimentation before attempting major modifications or dramatic changes.

SECONDARY DEVELOPMENTS

Building on this early leadership and example there soon followed a second wave of instructional planning systems. For example Abt Associates, of Cambridge, Massachusetts, were quick to establish a reputation in the urban and regional planning field through the procedures evolved for their NORTH-EAST CORRIDOR TRANSPORTATION GAME (Abt 1967*a*). This model was commissioned by the National Bureau of Standards to provide an environment for investigating some of the social, political and economic factors which influence regional transportational planning in the 'Corridor' between Boston and Washington D.C. More particularly, Abt's game sets out to provide a structure in which government officials can explore the negotiating processes followed by regional representatives in pursuit of transportation planning objectives. At the same time, the simulation procedure aims to broaden the participants' understanding of the comprehensive planning process at the inter-state level.

The CORRIDOR model is of particular interest to this historical review as it represents one of the roots of an Imperial College instructional simulation known as THE LAND USE–TRANSPORTATION SIMULATION which was designed, very largely, by Macunovitch (1967), a former member of Abt's 'Corridor' research team. Professor C. D. Buchanan, on behalf of the Transport section at Imperial College, commissioned the P.T.R.C. consultative group to design this game as a basis for a series of simulation exercises for use with his planning and traffic engineering students. The resultant 'human player' simulation creates a framework in which 13 to 36 players can experience aspects of the urban planning process and hopefully begin to appreciate some of the cause and effect relationships attendant upon alternative community decision-making chains. The setting of the simulation is a fictitious town with a population of 50,000. Some participants play such local authority roles as: planning officer, engineer, education officer, housing manager, treasurer and transport manager, whilst others play roles representing industry, wholesaling, retailing, construction, land development and various segments of the citizenry. Three 'cycles' constitute a complete game and each cycle represents a five-year period which, in turn, is composed of three phases. In the first phase, commencing with 1950 data, the players negotiate and make decisions appropriate to their individual and community needs or aspirations. In the second phase, the non-official players assume the roles of councillors and take local authority decisions. In the final phase of each cycle, all the participants co-operate to evaluate the consequences of their actions thus far, in order to define the starting point for the next five-year cycle. To simulate an entire 15-year-time span, at least two complete days are needed and present experience suggests that anything up to a week might reasonably be devoted to the exercise. The game is still in its developmental stages and will doubtless profit from considerable refinement in the wider trials now made possible by the availability of a complete LAND USE–TRANSPORTATION SIMULATION Kit (P.T.R.C. 1968).

A number of other games have come from the pioneering designers in the field and their innovative models have in turn also spawned related research efforts. Hendricks, the designer of POGE, has built the S.F.C.R.P. MICRO MODEL as a manual operational game to train San Francisco planners how to operate a computerized model of the City housing market which was developed in the office of Arthur D. Little by Ira Robinson, Martin Ernst, Francis Hendricks, Robert Barranger and Claude Gruen in 1963. In addition, Hendricks has been involved with a University of Pittsburgh team in developing the G.S.P.I.A. MANAGEMENT PLANNING GAME – a research project which was initiated in 1966 and is still under development (Hendricks *et al.* 1966).

Meanwhile, the basic CLUG model devised by Feldt has been continuously revised and upgraded by the originator and his associates. In particular Feldt, in association with others, has been involved in formulating a number of fairly finely calibrated models representing real-world regional settings. Amongst these models are: a modification of CLUG to show the impact of urban sprawl on agricultural land; an eleven-county regional game of Central New York State; and a three-county simulation of Washington D.C. Metropolitan Area. In connection with the latter project, the Washington Center for Metropolitan Studies commissioned Feldt to help to modify CLUG to the special topographical, industrial and residential character of the United States Capital. As a consequence, the Washington Centre and Environmetrics of Washington are developing and using several gaming simulations, such as CITY I and REGION (Urban Systems 1968*a* and 1969*b*) to give a wide range of people a greater understanding of the various ways cities are both changing and growing (House and Patterson 1969).

Simultaneous with the preceding American outgrowths from the Cornell game, Anderson (undated) at Michigan State University's Urban and Regional Institute has produced a programmed introduction to CLUG composed of 80 carefully staged 35 mm slides, a tape-recorded commentary and a super 8 loop film. In addition, a basic kit of CLUG materials has been mass produced by Dotson and Sawicki (1968) and is commercially available complete with a Fortran IV computer programme. Meanwhile, in Europe, CLUG developments have been less ambitious; for example Taylor and Carter (1967) have experimented with a series of minor modifications for undergraduate planners, and Taylor and Maddison (1968) have been concerned with an anglicized derivative, LUGS (LAND USE GAMING SIMULATION) as a demonstrational prototype. This, in turn, has spawned a further English development (Domitriou 1971), a Dutch derivative (Van der Heijden 1968), and a Swiss elaboration complete with a sophisticated referenda component (Taylor and Geiger 1968). More recently, at the American/Yugoslavia Project in Ljubljana, one of Feldt's former students has evolved a NEW TOWN game (Lawson 1968) which during 1968 was used experimentally as a sub-model to Cornell's basic COMMUNITY LAND USE GAME and since then kits of this game have become commercially available.

A final example of gaming research stemming from Feldt's work can be found in the University of Birmingham's local government postgraduate training programme. The Institute of Local Government's initial LOCAL AUTHORITY GAME (ILAG) was developed by Armstrong (1968) and, although acknowledging CLUG and METROPOLIS antecedents, is very much a product of local needs. Whatever the pioneering linkages involved, Armstrong and Hobson (1969*a*, 1969*b*, 1969*c* and 1969*d*) have gone on to much refine their gaming activity in response to developing course requirements and outside consumer demands. As a consequence, the Institute now has a ' nest ' of interlocking gaming models which is a tribute to their designer's ingenuity and the wealth of simulation ideas springing up around them.

Duke's METROPOLIS, rather like Feldt's CLUG, has been through a comparable series of elaborations and adaptations. In Germany, in particular, a number of these developments has been directly assisted by Duke, and the work of the University of Berlin in this direction has been described by Schran (1968) and Geiger *et al.* (1968). Two of the most unusual outgrowths from Duke's initial conception have been COMEXOPOLIS (Reske *et al.* 1966) and APEX (personal communication); these studies are designed to develop a series of computerized training exercises for administrators in the field of air pollution which have been supported by the University of Southern California and the United States Department of Health, Education and Welfare, with Duke acting as a senior consultant, and his Environmental Simulation Laboratory serving as design advisers.

Perhaps of greater significance to this discussion is METRO (MICHIGAN EFFECTUATION, TRAINING AND RESEARCH OPERATION), an ambitious gaming-simulation project currently being developed in the Environmental Simulation Laboratory of the University of Michigan under Duke's direction (Duke *et al.* 1966, Ray and Duke 1967). The major aim here is the production of a pedagogic device for training urban planners, politicians and related social scientists as well as possibly educating members of the public. This project, the implementing agency and its director are of such importance to the development of instructional simulation systems and planning games in particular that before continuing this historical survey a few words on this work in progress would seem relevant at this stage. Such elaboration seems to be called for if only because of the expertise involved, the substantial financial backing received and the fact that, although METRO is an extremely elaborate conception, all the models and sub-routines so far devised appear very adaptable and in most cases are thought by the designers to be replaceable by alternative formulations and this means that the game shows every likelihood of providing a useful test-bed for a diverse range of experimentation.

In essence, the METRO model is designed to create a miniature world which is representative of the typical, middle-sized American metropolitan area; demographic, political and socio-economic information for the model being drawn from Lansing, Michigan, during the period 1960–5. As can be seen in Figure 3, four kinds of roles are prescribed (politician, planner-administrator, educational administrator and land developer) and players representing these interests are located in each of three political jurisdictions (a central city district facing growing problems of congestion and blight, an upper- and middle-class suburb, and a zone of scattered townships just beginning to feel the effects of urbanization). Thus players have to grapple with both vocational and area interests at one and the same time. The players are presented with guidelines for action but, as in real life, they are under no obligation to follow this guidance.

The complete gaming vehicle relies upon a family of models and a battery of support facilities of which the major components are:

(a) a growth model which generates the aggregate pattern of growth in terms of population and exporting industries;
(b) an ecological redistribution model which reallocates households and firms; this is an elaboration of the TOMM Model (TIME ORIENTED

Figure 3. METRO *functional interactions*

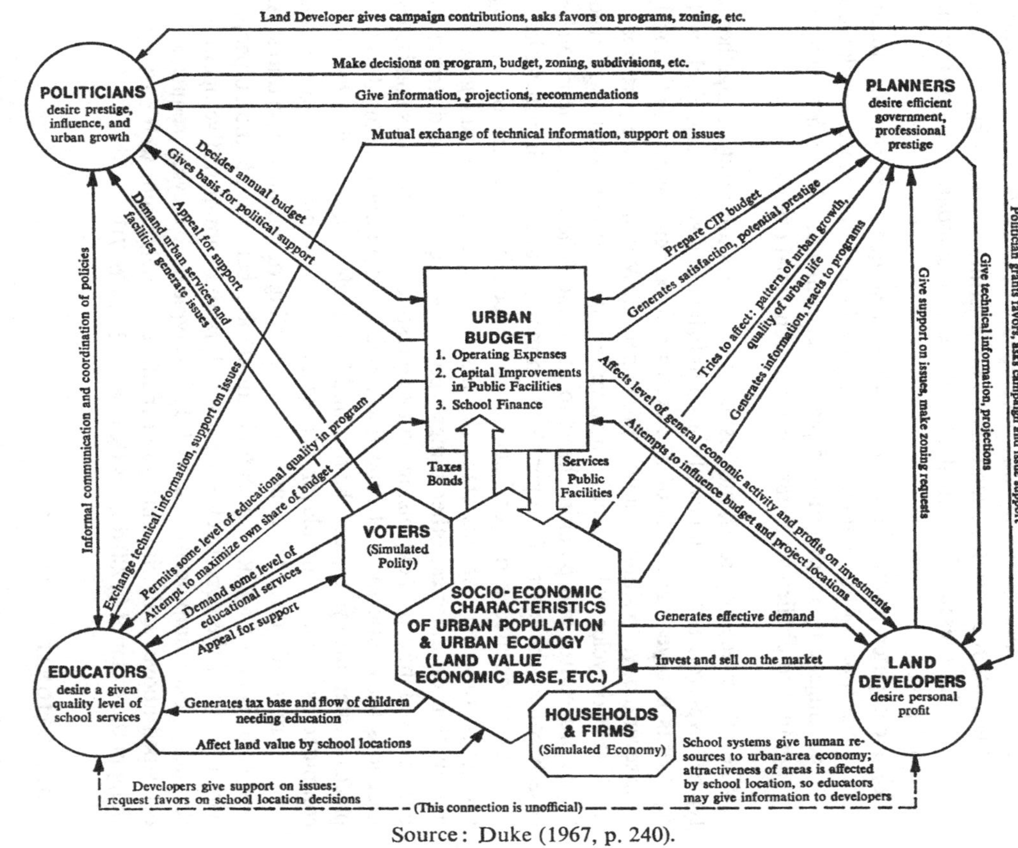

Source: Duke (1967, p. 240).

METROPOLITAN MODEL) originally designed by Crecine (1965 and 1968) during his earlier association with the Pittsburgh Planning Department;

(c) a voter response model which uses a 'Monte-Carlo technique' to simulate polling data and public response rates to both issues and candidates;

(d) a computer mapping technique called SYMAP (SYNAGRAPHIC COMPUTER MAPPING PROGRAM) originally developed by Fisher at Harvard. In brief this system translates such data as birth, crime and building rates into pictorial presentations, in shades of grey, on standard computer print-out paper.

(e) a computer printed newspaper, whose headlines alert players to the vicissitudes of the changing economic and political climate, very largely contingent upon patterns of player performance plus a few pre-programmed challenges stemming from archetypal events.

When the preceding roles, models and techniques are brought together, a typical five-cycle METRO run can be accomplished in a single day. Each cycle represents one year and sets of five cycles can be run almost continuously over several days to simulate any required period. With neophyte participants the first two cycles will normally involve familiarization sequences relevant to the procedures and planning basic to the game. The third cycle in such a case becomes the start of proper decision-making and training in which a carefully elaborated series of issues is developed in fulfilment of METRO's pedagogic function. At this stage the first actual computer run generates information on amongst other things: population growth and distribution, business and industrial expansion, and vote response to candidates and issues. Elections for politicians and education administrators are scheduled to take place on alternate cycles, and the fifth and final cycle, in a set of five, generates computer assessments of player performance and community growth, and these form material for the concluding de-briefing or critique procedures. As now conceived the METRO simulation can involve between 25 and 30 people, with 6 or 10 of the personnel being instructors and administrative operators. Running this computerized model for a typical five-cycle sequence, as described above, and with this number of personnel, might well cost something in the order of 2,000 dollars (Ray *et al.* 1966, p. 7)!

INSTITUTIONAL INITIATIVES

Before getting involved prematurely with matters of cost there is one group of models which has yet to be considered in order to complete this survey of planning games. This gaming group might be viewed as essentially 'institutional' or 'local' products in the sense that they very much reflect a special agency's needs and are often subjected to only limited use by the developing agency and closely related associates, largely because of restricted documentation.

For example, outside academic institutions several exploratory gaming models have been designed for particular practice situations. In 1966 Mitchell developed a game called URBAN PLANNING SIMULATION as an instrument for culling user information (Berkeley 1968, p. 62). The game has been expanded by some of Mitchell's students at the Harvard Graduate School of Design and has been played by citizen groups at a variety of socio-economic levels. As a data-gathering device player actions are seen as replacements for what the participant might find it difficult to verbalize when asked to consider the determination and evaluation of personal as well as community needs and desires. As presently conceived, the game can operate at any spatial level and basically provides no more than a rudimentary framework for multi-person confrontation on urban affairs issues (Mitchell, undated).

A similar form of game known as TRADE-OFF has been constructed by Berger and Walford (Berkeley, 1968). Here the designers also view their game as a data collection device but, in addition, stress its educational potential in the extramural area. In TRADE-OFF, players are asked to improve their community with a prescribed budget. Point values are attached to particular development projects and it is left to the players to determine priorities on a scale that inter-relates each capital project. Thus as dollars are spent, points are accumulated. Once participants have disposed of their allotted sum and a ranked list of improvements has been determined, they are then required to reconsider their previous decision-making cycle with the same objectives in mind but this time with a smaller financial allowance. In essence the game is a priority allocation procedure involving a ' trade-off ' process between alternative improvements and related costs. The game has been used in a variety of indoor and outdoor settings and Bourgeois (1969, pp. 116–17), the Director of the St Louis Model City Agency, has been particularly enthusiastic about the role of such community games as tools for facilitating ' grass-roots ' involvement and understanding.

On a rather different plane and this time springing from the university environment is the TORONTO GAME (Benjamin 1968). Although this model is no longer used by the designers either in its original form or within the initial educational context it does appear to be typical of what might be called ' in-house ' or ' one-off ' games and is also of interest because it has served as the source for a number of related research efforts (Benjamin 1969, personal communication). The original version of the Toronto model (TOG) is a simulation centred on Don Mills, Ontario, and the basic components of the exercise are identified in Figure 4. Structurally, the game incorporates many of the standard features of planning games which have been previously discussed; however, as the educational level and objectives are more fully identified than most ' institutional ' games, these are worthy of restatement.

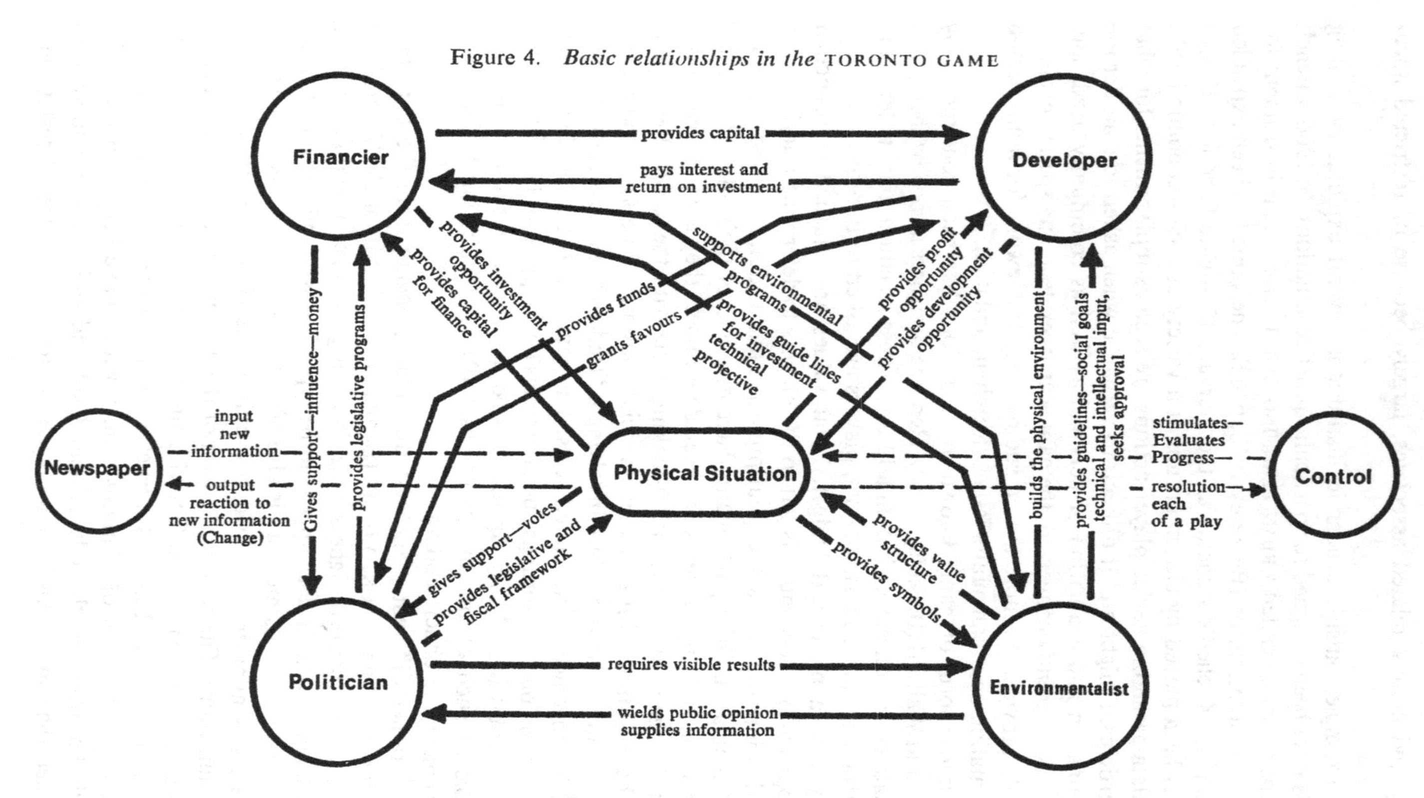

Figure 4. *Basic relationships in the* TORONTO GAME

Source: Benjamin (1968, p. 527).

In brief, the TOG model was conceived and used as part of the fourth-year design curriculum in the School of Architecture at the University of Toronto to satisfy five specific purposes: first, to introduce students to the interplay amongst many of the major elements of the urban system; second, to develop a method for structuring information, in which values and constraints are explicit rather than implicit; third, to highlight specific alternative value systems; fourth, to compare and integrate the development of game processes with the evolution of urban design strategies; and finally, fifth, to examine the validity of design alternatives through gaming. On this last point, Benjamin (1968, p. 528) has been particularly encouraging, in placing considerable confidence in the short-term value of games as *ad hoc* alternatives to the traditional architectural design juries by providing a different basis for a positive exchange of ideas and an exploration of alternatives.

One distinct branch of the 'institutional game group' encompasses a number of models that collectively involve varying degrees of staff–student commitment. Just as Feldt's CLUG I originated out of a class project, as it happens in an urban ecology course, so a number of other games have emerged from similar assignments in related urban studies courses. The documentation and utilization of these models vary greatly, as does the staff or student responsibility, but as yet none appear to have been used or publicized as widely as CLUG and hence remain in the category now under discussion.

Perhaps one of the best documented of these 'in-house' efforts is the Harvard Graduate School of Design's URBANIZATION AND CHANGE MODEL (Steinitz and Rogers 1968). Here an experiment in interdisciplinary education was channelled into the building and operation of two distinct yet related types of model. Using the south-western sector of the Boston City Region as the study area, five allocation models (industrial, residential, recreation and open space, commercial centres, and transportation) were developed alongside four evaluation models (local politics, town finances and taxation, visual quality, and pollution). The resolved sum of the models was a land use allocation plan considered in five-year stages to build up a twenty-five year simulation of urban growth, first as a projection of current development attitudes and, second, as an estimate of developments that might occur if a single form of metropolitan government could be established. These simulation runs were spread over sixteen weeks and the resultant 'game' was viewed as an experimental vehicle for synthesizing into a coherent planning model the analytical abilities of students and staff from various disciplines. Now the methodology has been laid down, similar 'games' can be more quickly accomplished. It remains to be seen whether the exercise is repeated at Harvard or whether others capitalize upon their well-documented experience.

Finally, when considering work in this direction, it is interesting to note what appears to be a further development in the use of instructional simulation to explore the possibilities of gaming as a vehicle for examining the viability and credibility of student projects (Henderson 1968). Independent of Benjamin's (1968) experimentation with a game as an adjudication device, which was described earlier, students in town and regional planning at Dundee have formulated a system of gaming-simulation procedures to consider the merits of regional strategies produced by other colleagues. The resultant game portrays the interaction between certain key regional settlements (Aberdeen, Inverurie, Huntly, Keith and Elgin) as well as between influential agents such as government and industry. From this interaction a growth competition model has been established with a points system built upon prescribed industrial, housing and amenity factors. The students very largely defined their own brief, drew up their own game rules and designed a revision procedure for amending constraints over time. Such has been the success of the operation, as adjudged by students and staff, that similar exercises have been repeated at Dundee and at least one other Scottish Planning School has adopted, and is developing, a comparable simulation framework (Parham 1969, personal communication).

Before concluding these remarks on what has been called 'institutional' games, it is important to realize that some of the models in this category have remained local products not entirely because of their limited relevance to other situations and other users, but possibly because of the early state of development work and the consequent lack of comprehensive documentation *or* because in certain cases, where institutional expertise is at stake, the developing agency has made it a matter of policy not to release material for unlimited use. Some examples of models restricted in both these ways are now considered.

Goodman's POLICY NEGOTIATION GAME sometimes known as the INFLUENCE ALLOCATION GAME is but one of the many models under development, at the present time, whose documentation is particularly sparse. However, the basic format of the model is now well established and the game has been widely demonstrated by its designer. The actual framework of the model owes much to Kaiser's FUTURE game and the elegant logistics and display techniques evolved by Helmer, Gordon and Goldschmidt (Taylor 1969*b*). This derivative of FUTURE has been used to demonstrate an open-ended approach to game design which is significant in its emphasis on the participants' modelling ability as opposed to their playing skill. Goodman has used the framework, very largely, to build up a confrontation simulation relevant to the educational situation in a particular Michigan community, but also has been quick to demonstrate the ease of building or adapting the game structure to very different situations. A good many instructional simulations have involved students in the model building process and in game modification procedures but here the aim is more specifically orientated towards establishing a means for allowing instantaneous re-norming *or* complete rebuilding to be undertaken as play evolves. In this way, the environment for learning can be adjusted in response to group pressures and individual requirements. The idea is certainly an attractive one and in demonstration very persuasive in the skilled hands of Dr Goodman; it remains to be seen whether this interesting development can be fully exploited by a wider audience.

An equally promising research effort is underway at the Drexel Institute of Technology in Philadelphia and, like Goodman's project, has yet to reach the state where the literature has caught up with even the preliminary experimentation. Under a 500,000 dollar National Aerospace Agency grant for 'Research and Education in the Technology of Management of Large Scale Programs', a community development simulation called BUILD is being designed (Orlando and Pennington 1969). BUILD is planned both as a mathematical model and as a role-playing computer game in order, amongst other things, to (Pennington 1969):

(a) enable improved public participation in the urban development process, whereby the citizen can contribute valid inputs to the process at a technically sophisticated level,
(b) advance better hypotheses about the functioning of the urban system,
(c) reveal and resolve some of the extremely complex psychological and social relationship problems of individuals involved in the planning process.

The central purpose, behind the study, being to demonstrate some of the applications of aero-space technology to urban and related social problems, with the present 'state of the art' in the development of instructional simulation systems such transference of expertise is to be welcomed and obviously further initiatives in this direction should be actively encouraged.

Finally, an example of a gaming model which is fully operational, well documented, and partly restricted in use is to be found in the ROUTE LOCATION GAME (Creighton *et al.* undated, and Frye 1970). This game has been designed by, and is the property of, Creighton Hamburg Incorporated – a group of American planning consultants – who have established a compensation rate for running the game to cover their development and demonstration costs. In the United States, for example, a day-long presentation for educational institutions costs 450 dollars plus demonstration staff expenses (Roger Creighton 1969, personal communication). So far this expenditure apparently has not deterred its use by several groups of urban planners and highway engineers who have been involved in running the model as part of training programmes in the United States Department of Transportation, in Washington D.C., and in the University of North Carolina at Chapel Hill.

As the name suggests, the ROUTE LOCATION GAME is concerned with the delineation of an alignment for a public highway. Three alternative intra-city routes are open to consideration and the game revolves around the public and private interests at stake in arriving at any locational decision. In essence, the exercise seeks to uncover, as well as to weight, various community values and aims to provide a greater appreciation of problems and decisions faced by those involved in the urban planning process. As with the majority of planning games, participants are expected to benefit from being given the opportunity to see a procession of events in a dynamic context, freed from what might best be termed ' considerable background noise '.

This professional exploitation of ' spill-over ' from normal consultancy practice has its counterpart in the field of educational commerce. A number of new companies or non-profit groups have been established recently to propagate and exploit gaming-simulation merchandise and expertise alongside other educational services (Taylor 1969*a*). Whereas in the past instructional simulation system consultants, such as Abt, have not entered into marketing projects, an obvious change is taking place, and more ' packaged ' commercial material is coming from simulation specialists. This material is pitched at various educational levels and as one might expect is extremely uneven in terms of its quality. In short, this is probably a growth area where already the products are too numerous to be covered here.

Thus, this section has presented selected examples of instructional simulation research stemming from practitioners, research workers and teachers or students with varying degrees of academic, professional or commercial motivation involved. These, and similar developments, will be drawn together in the ensuing chapter on the dissemination and educational features of such models, but before making any comments on issues of this kind it must be appreciated that there are many other 'discipline orientated' games employed in a variety of urban and regional studies programmes. It is such models that are the concern of the final section in this chapter.

SOME RELATED MODELS

The foregoing historical survey has endeavoured to confine its attention to particular forms of instructional simulation more specifically relevant to the training of professional planners. In a field as ill-defined as planning, the writer would not wish to exclude from this account any work which although concerned with some aspect of the urban development process might fall outside narrowly defined professional terms of reference. For example, the under-developed ' area ' applications of, and the building and social science approaches to, simulation are obviously of some relevance here but are not something to be tackled comprehensively in this publication. However, to conclude this chapter in a more complete manner, it might be considered worth-while to indicate briefly some urban and regional affairs games which come within these categories and are used more widely outside planning education.

Examples of instructional simulation systems applied to less- or under-developed situations overseas are few in number and, because of their scarcity, these systems are now considered in somewhat greater length than the building and social science applications.

THE VIRGIN ISLANDS GAME, sponsored by the College of the Virgin Islands under the auspices of 'Title I' of the United States Education Act, was designed by Abt Associates (1968). It aims to clarify some of the internal communication patterns amongst government and the community in a developing nation. The medium for exploring these inter-related networks is a simulation of public safety, housing and recreation planning in the town of St Thomas. Basic statistical data on the Island is provided, and over twenty profiles of various interest groups and decision-makers are described. The underlying theme of the game is an enactment of human involvement in the process of community development over a five-year period. During this simulated time span the model acts as a loosely structured framework for players' discussion of the more pressing issues likely to affect a developing township. For example: How should funds be allocated to meet the needs of a growing community? How should the expanding community meet the conflicting demands of more nationals, aliens and tourists? In effect, the operational game brings together up to fifty people to think through community problems via an elaborately controlled 'teach-in' procedure designed to reveal opportunities for improving communications amongst participants representative of the Islands' residents and non-residents.

Of particular interest to this brief survey is the game's specific orientation, its wider applicability and its inherent flexibility. Although the 'scenarios' concentrate on issues likely to affect St Thomas' growth, Abt (1968, p. 4) has pointed out that a change in subject-matter would not require any major reconstruction of the basic format. For example, some of the profiles would have to be amended and a few issues added or removed, as appropriate, if the game were used for an alternative Virgin Islands township. Similarly this form of multi-person 'scenario' simulation could readily be centred on other related problem areas inside or outside the Islands. Such an approach is particularly valuable because of its ability to involve, actively, large numbers of the public in the local government planning and decision-taking process. Abt's model has done this in two ways: first, by getting a large cross-section of the community to contribute toward the construction of the game itself; second, by facilitating extensive participation in the operational game from many representatives of diverse interest groups. Each run of the game can equally well create a 'teaching' or 'learning' environment in the sense that students and professionals, with differing objectives, can participate together. In this way, the students would be receiving *tuition* on how government decisions are or can be made and more experienced members of the community, whilst bringing added richness to the game through their own experience, would be free to *learn* more about new ideas and fresh solutions to problems which in all probability run counter to their everyday real life position and views. Thus, through cross-sectional participation and role reversal procedures an increased understanding of other people's attitudes and actions concerned with planning policies can be achieved at modest cost and little inconvenience. The ' pay-off ' for both game designer and participant is the extent to which the simulation offers more by way of communication, than ordinary public discussion or academic debate.

A totally different, yet equally flexible, instructional simulation of a developing country situation is to be found in Co-operative Educational Services' SIERRA LEONE and SUMERIAN games (Wing 1968). These two computer-based ' one-person ' games are essentially school exercises in which the student plays ' against ' a computer program. Whilst the educational level of these exercises relates to areas outside the scope of this work they merit a slight digression as it is possible to view such models as stepping-stones to more advanced simulations being developed elsewhere, and Bell's (1970) research in this direction will be discussed shortly.

For the moment, the simple and straightforward mechanics of these computer-based approaches will be described, briefly, as they operate at the school level. Economic problems of a newly emerging nation, within a specific cultural and geographic setting, are presented via typewriter display and projected slides. In the SIERRA LEONE case: first, the student must pass a preliminary examination on the required geographical and economic background; second, upon successful completion of this entrance test, the student assumes the role of a junior economic advisor and is confronted with a series of decision opportunities; third, and last, as the student progresses from problem to problem and satisfactorily demonstrates his understanding and problem-solving ability so he is promoted and consequently is given tasks of increasing magnitude and difficulty.

This programmed escalation of academic challenge is not as common as one might expect in gaming. However, the man–machine interface is an attractive and popular research area, and a distantly related elaboration of the preceding form of simulation is the DEVELOPMENT GAME designed by Clive Bell (1970) for use in the Institute of Development Studies, at the University of Sussex. This man–machine simulation exercise in economic decision-making is structured around a mathematical model of a hypothetical ' less-developed ' economy. Prescribed areas of production have outputs acting in proportion to the levels of investment and the direct proportional relationship is modified, from time to time, by stochastic elements representative of such vagaries as ' Acts of God ' or industrial ' teething troubles '. Players have certain variables they may manipulate; consequently the activity levels of certain production areas are determined by player-selected labour inputs, and similar questions of choice determine, amongst other things, domestic and trade taxes, price and stock exchange policies and current rates of interest.

The normal game comprises some ten to twelve cycles representative of an equivalent number of years, and at Sussex a run of this length has been spread over two or three days. No single set of outcomes is identified as the product of an optimum strategy but success is gauged with reference to the achievement of initially stated player objectives modified by an assessment of the related implementation difficulties. Consequently each team is ' marked ' according to the weights they attached initially to growth, inflation, balance of payments and distribution of income objectives. Bell (1970) has indicated that he expects that the best way of using the game and the lessons to be drawn from it will emerge through experience rather than *a priori* argument. In particular, he has cited (1970, p. 132) two functions for the simulation:

> In the case of civil servants . . . the main purpose is to teach that variables *are* inter-related, rather than *how* they are inter-related. Practical men often work, explicitly or more commonly implicitly, with very simple and occasionally uni-causal models. The purpose of this exercise is to wean them from such intellectual habits and to open their minds to the need for the analysis of complex systems. The aim is more in the nature of a therapy for intellectual cramps than the presentation of a universal model . . . The scope for students and professional economists is rather greater. Freed from the limbo imposed by *ceteris paribus* and *mutatis mutandis* conditions which are often a feature of economic analysis, they can seek to use their analytical tools in order to ' optimize ' the performance of a general equilibrium system. Moreover, the programme need not be used in the ' play ' mode alone. It can also be used as the basis for exercises in comparative statics or parametric sensitivity.

All of these developing country applications are extremely recent and workers in this area are *especially* careful to point out: their limited experience; the speculative nature of their findings; and the great need for much more testing and revision. At the same time, initial reports have been sufficient to encourage further efforts in this direction and to merit, in the writer's opinion, inclusion in this short review of gaming-simulation activity outside the main stream of professional planning applications.

In the building sciences, simulation exercises are more common and there is a growing variety of games concerned with the background to, and the implementation of, planning policies (Negroponte 1970). OPERATION SUBURBIA (Zoll 1966) and LOW BIDDER (Entelek (1964) are but two of the relatively simple ' package ' games designed to simulate, respectively, conditions encountered in the land acquisition and construction tendering process. Webb and Wheeler (1962) have designed OPERATION TAURUS as a more sophisticated executive level management exercise on the economic aspects of building and civil engineering; this four-company model was commissioned by Howard Farrow Ltd, and John Laing's Ltd (undated) have their own company counterpart in EXERCISE QUINTAIN. The complete design and building process is simulated in FACTOPLAN (Polycon 1967) which is a training programme relying heavily upon a development project game devised to include all the major members of the building team. An example of a similar approach with comparable objectives is to be found in the Ministry of Public Building and Works EXTRACOL syndicate or collaborative project (Cooke 1970). Finally, Cohn's (1968) ARCHITECTURAL CONTROL SIMULATE must serve as an example of one of the few games in the building sciences which focus upon a particular component of the development cycle. Here, interest rests upon aesthetic aspects of development control and the machinery for handling case study planning applications.

Social science gaming experience in urban and regional studies is perhaps even more extensive than the sum of the development area and building science activity. At the global level THE WORLD GAME (Fuller 1967 and McHale 1967) and FUTURE (Taylor 1969*b*) are two of the ambitious projects conceived as a means of establishing an overview of international trends. At a sub-regional level, Goodman's (1968) INTER-COMMUNITY SIMULATION and Enzer *et al.*'s (1969) STATE POLICIES GAME are recent examples of experimental approaches to some socio-political aspects of regional economic planning. SIMSOC (Gamson 1966 and 1969) focuses on urban social order problems through examining the processes of social conflict and control. SIMPOLIS (Abt 1967), PLANS (Boguslaw *et al.* 1966) SITTE (Western Behavioral Sciences Institute 1969) and FEMENTO (Meier 1968) are but four examples of the many urban society games involving national and regional interest blocks as well as community pressure groups. At a local level, Davison's (1961) PUBLIC OPINION GAME simulating township and neighbourhood attitudes is an exercise designed to throw light on factors involved in opinion formation and political participation within smaller urban settlements. Finally, at an institutional level there is a series of games which simulate organizational systems and these models include: THE EDUCATION SYSTEM PLANNING GAME (Abt 1965); THE COLLEGE & UNIVERSITY PLANNING GAME (Forbes 1963); THE MADINGLEY GAME – a management exercise for the public sector of higher education (Piper and Rae 1969); and THE HOSPITAL EXERCISE IN LONG TERM PLANNING (P.E. Consulting Group 1966).

The number and variety of games in these three fields, which overlap with planning, varies very much from subject to subject, and obviously the preceding notes are hardly an exhaustive treatment of all the areas where games are thought to be useful or relevant. The main object in presenting this material has been to identify some related yet, in this context, somewhat peripheral applications of instructional simulation before going on to consider the impact and possible educational utility of the technique with reference to planners' approaches to the urban development process.

5. Impact, pitfalls and pay-offs

Prompted by increasing and changing demands placed upon planning education and the emergence of a promising new instructional technique, the writer set out to examine a gaming-simulation approach to the urban development process. Since it is believed that gaming models have some pedagogic potential for planners and yet for almost a decade have lacked anything by way of comprehensive assessment, this chapter seeks to establish a clearer mental perspective on the nature of the technique in relation to the extent of its impact and potential usefulness.

THE LEVEL OF COMMITMENT

Few stock-taking assessments of the employment of instructional simulation systems in differing subject areas are known to the writer. Without exception, all commitment appraisals appear to have dealt with the field of business and management studies. Here the surveys of Dale and Klasson (1964), drawing upon 90 informants, and Shubik (1968), relying upon 48 respondents, provide useful insights into the nature of business gaming, its development, dissemination and utilization. The only comparable work undertaken in relation to urban studies programmes appears to be that recently carried out by the writer (Taylor 1969*a*) relying upon 70 questionnaire responses received, very largely, from European and North American architectural and planning schools. In the paragraphs which follow, some of the findings from this utilization survey and the resulting directory of urban development games are summarized as a preface to brief descriptions of more localized validation studies to be found in the literature.

The 1969 survey revealed that 46 out of the 70 institutions questioned used instructional simulation techniques in some way or other, although in almost every case their specific level of commitment was extremely unclear. The rather sketchy material collected by way of estimates of time and resources devoted to gaming-simulation in each of the 46 institutions is well illustrated by Table 2. CLUG was easily the most popular model and was used by 16 institutions, with CITY I, METRO as well as P.T.R.C.'s LAND USE–TRANSPORTATION SIMULATION each reported to be used by six different agencies. It is interesting to note that all these models are particularly well documented in terms of published material and CLUG and P.T.R.C.'s game have the added advantage of being available in commercial kits.

Table 2. Estimates of comparative costs for various gaming procedures

No. of institutions who have used or are using model	Name of model	Capital cost estimate	Design labour commitment	Running cost	Current package material price
6	CITY I	$150,000 [1]	25 people	—	—
6	METRO	$500,000 [2]	30–50 man yrs.	$2,000 per day [3]	—
1	DEVELOPMENT GAME	—	1 man yr.	—	—
39	CLUG	$73,000 [4]	4–5 man yrs.	—	$125
6	LAND USE–TRANSPORTATION SIMULATION	£2,200 [5]	$\frac{1}{2}$ man yr.	20 gns. per day for a prof. adviser	£95
3	ILAG	£150 approx. exclusive of labour costs	$\frac{1}{3}$ man yr.	—	—
4	LUGS	£65 approx. exclusive of labour costs for principal investigator	$\frac{1}{6}$ man yr.	—	—
3	ROUTE ALLOCATION GAME	—	—	£450 plus travel expenses of a prof. adviser	£200 per session

NOTES:

[1] Inclusive of a $70,000 US Office of Education grant.
[2] Inclusive of Ford Foundation, Nat. Science Foundation, Dept. Housing and Urban Development, Dept. Health, Education and Welfare grants.
[3] Ray *et al.* 1966, p. 7.
[4] Total costs met by a $73,000 Ford Foundation grant.
[5] Inclusive of a £900 Nuffield grant in support of the Imperial College Contract.

Source: Taylor, 1969*a*, p. 114.

The 24 uncommitted ' non-user ' institutions, i.e. those centres not using gaming-simulations at the time of the survey, indicated that:

(a) lack of information and/or funds had deterred them (17 mentions);
(b) lack of willing or available staff had restricted experimentation (13 mentions);
(c) a lack of appropriate models had prevented their use (12 mentions);
(d) a lack of validation studies had deterred them (9 mentions); and
(e) two quoted administrative reasons which had restricted scheduling opportunities.

In addition, four respondents stated that they did not think that the technique was suitable to their needs and two of these expounded upon their personal dissatisfaction with regard to the spread of ' American methods '. However, on a more hopeful note, 20 out of the 24 non-user institutions indicated that they were *in the process* of considering gaming-simulation systems for future use.

A by-product of the survey was a listing of 62 urban models which were catalogued in the writer's Directory of Urban Development Games (Taylor

Figure 5. *The build-up of urban development games and the number of computerized models as identified by Taylor* (1969*a*, p. 230)

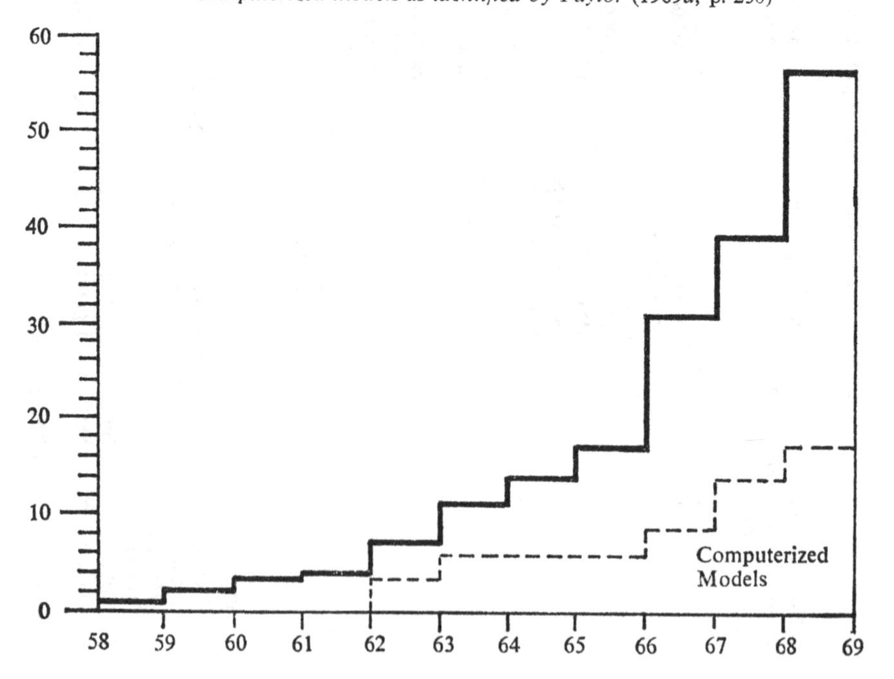

1969*a*). As the technique originated and, to date, has been largely developed in the United States, it was not surprising that over 90% of these models were of North American origin and specifically related to the American environment. Well over two-thirds of the games were designed after 1965 with both manual and computerized models enjoying a parallel rise in popularity.

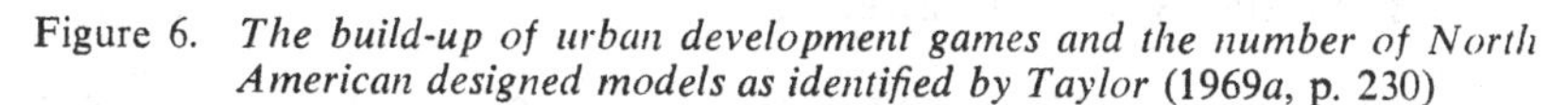

Figure 6. *The build-up of urban development games and the number of North American designed models as identified by Taylor* (1969*a*, p. 230)

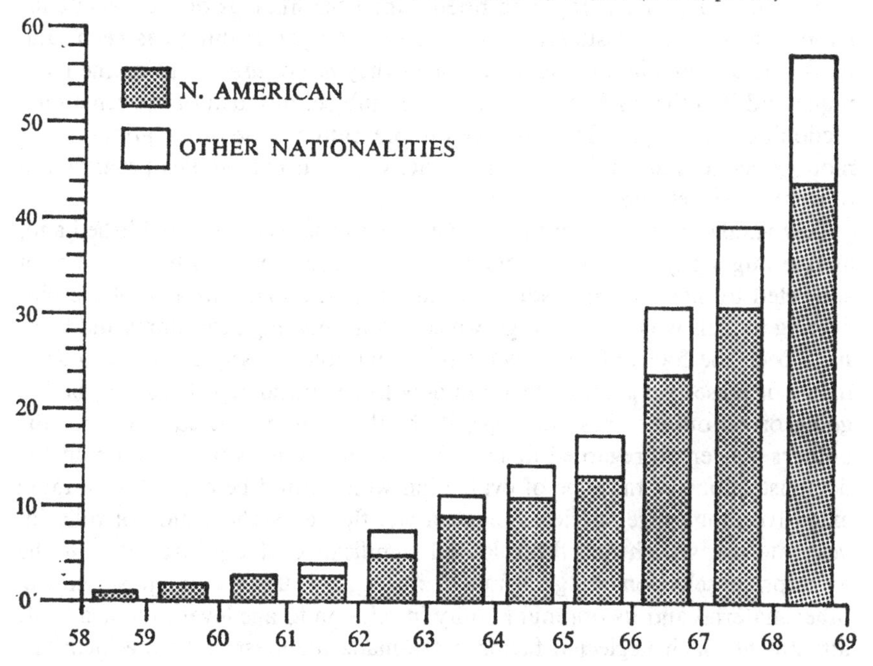

In terms of operational logistics: well over half the models in the directory required two or more administrators; 50% of the models involved a minimum commitment of over five hours and almost half of the games documented were designed for over 17 participants. The survey established that it was the institutions with smaller numbers of staff and more students which were shouldering the biggest burden with respect to gaming-simulation usage. Why this should be the case was not at all clear as the paucity of data collected prevented any correlation between the characteristics of available games and particular institutional resources. Hence, to explain the clear relationship between high staff–student ratio schools and instructional planning systems must rest, for the moment, with the writer's earlier conjecture: i.e., Is it that the games so far designed have in mind the larger audience and are not readily transposed for smaller groups? Are the bigger institutions with smaller staffs looking for time-consuming assignments for larger student groups? It is possible that institutions with a high staff–student ratio are more pressed to examine new techniques offering possible 'pay-offs' in relation to their need to improve their overall efficiency?

A SUMMARY OF INITIAL FEEDBACK

So far, this chapter has reported briefly upon the findings of the writer's instructional simulation survey (Taylor 1969*a*). As yet, nothing has been said, in any detail, on why instructors consider they *are* or *are not* using the technique and it is these reasons which are the subject of much of the remaining section of this chapter. However, before presenting these observations, some more concrete assessments of gaming activity in urban studies programmes deserve to be mentioned.

Few validation studies of planning games in use have been published and, as one might expect, it is the older, more mature models, which have been subjected to more detailed study. METROPOLIS and CLUG are two of the pioneering models which have outgrown constant updating procedures and these have been the focus of some evaluation attention; consequently it is assessments of these two games which are now to be considered. However, before going on to outline these findings, it should be remembered that both the authors concerned regarded their evaluation attempts as pilot studies and as demonstrations of the type of evaluation which could be employed in more extensive long-range studies. For example, the researchers did not fully investigate areas such as: the role and significance of the instructor or the appropriate selection of a game model, its length of use, its juxtaposition with other material and its optimum utility in relation to age levels and academic attainment. Such neglected factors were made manifest and were identified as obvious areas for further research.

In the METROPOLIS study (Duke 1964), the performance of two gaming groups was compared with that of a control group. In all cases the students concerned were Michigan State University 'seniors' enrolled in a terminal course on city and regional plan design. After testing to establish a base line, 30 students were separated into the three groups; two of the gaming groups were given a three-hour briefing and a nine-hour exposure to METROPOLIS. Meanwhile, the ten members of the control group maintained the 'normal' lecture and studio work pattern which had been established for the course. Within a 14-day period all groups were given a before and after examination; both groups which played the game showed a marked improvement in examination results as opposed to a small improvement for the control team.

In addition, approximately one month after the terminal examination was given, an attitude questionnaire was distributed to reveal all the groups' reactions to METROPOLIS. All replies were anonymous and students were encouraged to be frank. The response to this questionnaire indicated that: three-quarters of the total class had a desire to play METROPOLIS at a future date; two-thirds of those who had already participated in the game wished to repeat the experience and 80% of those in this category considered their commitment in the experiment worth the time and effort.

A more recent and more detailed study of some of the educational aspects of another pioneering planning game has been provided by Monroe (1968). Using the CLUG I model with three experimental groups and lecture course presentations with three control groups, she has compared the performance of these six matched groups of students in a Human Ecology course at Cornell University in the autumn of 1967. Although limitations of time and numbers involved in the investigation restrict any generalizations from the findings, this research does provide some interesting insights into CLUG's teaching effectiveness and potential usefulness as well as providing a case study of a commendable methodology for testing planning games.

The experimental design of this pilot study was concerned primarily with a test of the hypothesis – will a gaming-simulation model (CLUG I) teach the same body of material as effectively as a lecture course with associated reading assignments, given equal amounts of time for instruction in both methods. To test this hypothesis, matched groups were evaluated as either CLUG players or as lecture and reading course students. Three types of learning and retention tests were given to the total sample, which numbered 32 at the outset and 24 at the completion. The evaluation comprised: a factual test given twice, once immediately after the experimental application and once at the end of the academic term; an opinion rating; and a continuous series of participant and observer assessments.

The results of this comparative study indicated at least three things. First, whilst the CLUG model can teach facts, it is not particularly effective for teaching factual or theoretical material. However the game did incorporate an additional form of learning which was testable in terms of the dynamics of systems and, more precisely, in terms of how theory is affected by practice. Here it is of interest to note that perhaps the most important aspects of the research for Monroe was the implication that games have the ability to introduce a new dimension to the world of facts and theory by adding the pragmatism of experience (Monroe 1968, p. ix). Second, confirming the observations commonly found in the literature, games motivate most students in a way not achieved by other teaching tools. Third, the nature of the learning which took place over time, and beyond the actual gaming sessions, indicated that there is a positive interaction between games and lecture material, and a sufficient correlation to emphasize that games should not stand alone but should be integrated with lectures to constitute a broader form of course presentation.

To add to these findings, one must turn to the more subjective assessments of why institutions or individuals consider they have or have not used instructional simulation systems. For the most part, this information was elicited through the writer's 1969 utilization survey and it is the respondents' reasoning which is now considered alongside the observations contained in the literature. However, before presenting this material, some general comments would seem to be in order.

First, the survey respondents' replies very largely appear to stem from a growing consensus of opinion rather than from well conducted validation studies. Second, although the writer has endeavoured to present, as fairly as possible, the reported advantages and limitations of gaming-simulation procedures, it should be remembered that, in common with a great number of respondents, the writer has invested heavily in the technique. Third, and following on from this point, it should be realized that such a presentation is inescapably tempered not only with the bias of personal experience in constructing, modifying and using games but also, to some extent, any survey in this field is contaminated by the impressionistic comments to be found in the literature or issuing from fellow researchers, users and observers. Thus, certain characteristics which have come to be associated with gaming-simulation techniques are taken at their face value and reported here, bearing in mind that it is not the purpose of this chapter to weigh the merit of these comments, merely to point out that they are made, and made with some frequency.

LIMITATIONS, DRAWBACKS AND DETERRENTS

Straightway, it must be acknowledged that there is considerable instructor resistance to the concept of ' gaming ' as a pedagogic activity in higher education. Few survey respondents or critics in the literature have been explicit on this point but it appears to be implied that gaming goes hand in hand with fun, or joy, in learning which *may* be acceptable at a school level but is seen to be not entirely compatible with the more rigorous labours of the higher echelons of the educational system. Possibly this teacher resistance emanates partly from: a lack of interest in innovation and change; a substantial adherence to accepted approaches; the initial difficulty of experimenting with the unknown without guidance or assistance – in sum, a shortage of mechanisms to ease the process of change; and finally, premature judgements reached after superficial contact with demonstration sessions or cursory perusals of the literature.

Some critics have claimed that, on the student side, problems of ' wayward commitment ' can stem from the considerable ' unreality ' attached to gaming situations (Raser 1969, p. 27). The 'novelty' and 'play' connotations are seen by some to work against *serious* participant commitment. Despite the apparent intense involvement and high engagement associated with operational sessions some observers obviously find it hard to accept that decision-makers can behave responsibly in simulations where the risk and danger are no more than 'paper' consequences. Such qualms are voiced regardless of the absence of hard evidence on the documentation of reckless or frivolous behaviour and ignoring the advantages, on occasions, of uninhibited experimentation of the type which breaks away from the conventional wisdom. There is reason to believe, however, that this is a problem which, in theory, could well exist but in the writer's experience it rarely appears to be made readily manifest in either the literature or in actual operating sessions. This is not to dispute that this disquiet may exist in some minds and hence it is an issue which warrants further discussion at a later stage.

Despite the fact that, by the end of 1968, there were in existence over fifty gaming-simulation models dealing with various aspects of the urban process, survey respondents listing drawbacks associated with the technique considered the limited supply of models as one of the major handicaps. Not only did they point out this shortage but also drew attention to the limited operational experience from which they could draw support and guidance. In the majority of cases, gaming experience was limited to a preponderance of short runs and some of this experience consisted of demonstrational or 'stunt' sessions, more concerned with introducing an operational procedure as opposed to the achievement of certain educational objectives. What is undoubtedly lacking is widespread and lengthy operational experience. The norm amongst the majority of users is an exceedingly modest commitment to the technique in terms of time devoted to the total learning system and in terms of the numbers of models being exploited by any one agency.

To further exacerbate this problem, a large number of the 'non-user' respondents to the survey believed that games appropriate to their needs were not available *or* their shortage of staff and resources was said to have inhibited any substantial commitment in such an experimental direction – especially as some planning games were reputed to require considerable resources to expedite initiation procedures and yet, at the same time, the technique lacked a sufficient record of proven worth. Such a situation becomes doubly daunting when it is recognized that participant learning or achievement is difficult to test or measure. Thus, potential users are not only inhibited by the absence of a record of proven worth but also deterred by thoughts of having to consider undertaking their own validatory assessments. Such studies require not only time and effort, but, as Taylor and Carter (1969*b*, p. 11) have pointed out, often lead the investigators into foreign educational and statistical areas which, in themselves, require considerable mental readjustment. And yet, in passing, it should be noted that such studies are likely to have a highly significant 'spin-off' in the form of a constructively critical approach to, amongst other things, educational values and the techniques of teaching.

Finally, with regard to the availability of games, there does appear to have been a tendency to concentrate much of the model building work on certain more readily quantifiable areas, and especially attractive aspects, of the urban scene. Today's models may not be representative of tomorrow's range, indeed it is to be hoped that greater diversity will emerge. Similarly it is to be hoped that the time now occupied by 'de-bugging' procedures will be shortened. Thus, the writer would accept that not only is the number of 'mature' gaming models limited but also that their focus has been, so far, rather narrow and restricted.

Several gaming-simulation critics have been quick to point out that the preparation time and effort behind a comparatively small component of the higher education timetable is probably formidable. At this stage, it is difficult to challenge such a criticism as few meaningful cost estimates for game construction as well as frequency and length of use in specific situations appear to be available. Table 2 provides a summary of the sparse information located in the form of estimates of: capital cost, labour involvement and operating expenses relevant to particular planning games and the number of institutions using them. It is to be regretted that the table is in no way comprehensive nor does it purport to represent a cross-section of the urban development models of possible interest to the planner. If anything, the table is a testimony to the designers' and users' reluctance to commit their involvement with instructional simulation systems into straightforward economics.

At least two leading authorities in the field have expressed opinions upon the comparative investment in gaming relative to all other urban modelling and, as any figures or assessments in this area are exceedingly difficult to find, these opinions appear worthy of restatement. Meier (1968, p. 2) has suggested that thus far about one tenth as much investigation has gone into gaming-simulation of urban situations as into computer simulation – in his opinion, primarily because the latter has been applied to massively financed metropolitan traffic and land use studies. On a similar level, Feldt (1967, p. 4) has argued that game design costs are a very small fraction of those encountered in the development of many urban models and probably no more than one or two man-years work has been invested in the majority of today's planning games. He has gone on to point out that the costs for a particular game depend, of course, upon the degree of detail required, the system it is intended to represent, and the amount of previous work upon which the game construction is based. Although a considerable commitment in terms of time and effort may, at present, be required to design and run a gaming exercise, it does appear that once a body of games is constructed and popularized then operating costs are likely to continuously decrease as experience increases. Again, this is a point the discussion will return to, but for the moment, it must be accepted that the yield, or ' pay-off ', commensurate with the effort expended in gaming activities is in few ways assured.

If costs are related to use, then possibly only the CLUG model has so far demonstrated its widespread applicability and potential high cost effectiveness. Obviously, information on all aspects of costing is very unsatisfactory – and a state of affairs unfortunately not uncommon in education as a whole and higher education in particular. Without such data, the construction of a game, in isolation, for a single purpose, will continue to be considered by some as something of a luxury. This point has been well made by Shubik (1968, p. 650) in relation to business gaming and *may* be equally applicable to urban studies gaming activities. Certainly this investigation has produced no evidence to refute such a statement.

One point relating to costs, however, does appear to be clear: the extra-ordinary demands placed upon accommodation, timetables and equipment by gaming sessions in comparison with other teaching methods more commonly used in planning education are exceptional. This often overlooked, yet very serious, difficulty varies very much from model to model yet the fact remains that very few games, relevant to urban and regional development, are available in a form which can be run adequately in the standard classroom and in the conventional lecture or ' studio ' period. In these circumstances, it is not surprising that games become extra-ordinary occurrences or isolated events in many cases relegated to summer schools, extra-mural courses or even extra-curricular evening activities. Events in this last category have undoubtedly proved very popular and, bearing in mind the fact that gaming procedures can be time-consuming, it probably says much for the motivational attractiveness of the technique that students readily commit their ' free ' time in this way; time that perhaps would not readily be committed to some other academic pursuits?

Finally, stemming out of the question of cost and user requirements are the problems associated with the selection of a mode of computation appropriate to particular needs. Considerable contention surrounds both manual and computerized approaches to the book-keeping and accounting procedures fundamental to gaming, and a choice in every case must rest upon the individual teaching objectives and intrinsic communication qualities of each separate instructional simulation system. This is to say that such an issue cannot properly be debated in such a context and ultimately it is for every teacher to resolve his own computation requirements on the basis of each and every game run related to his setting.

On an entirely different level, at least one critic has voiced certain reservations about planning games in terms of their inflexibility and emphasis. For example, in a recent Town Planning Institute *Journal* article which constitutes a criticism of an earlier special issue of the *Journal* on new planning technologies, Kriesis (1968, pp. 226–7) comments:

> The moral of this exposition (Taylor and Carter 1967) is that the 'real world' simulation in the games proposed implies the acceptance of the *status quo*. A *status quo* of all aspects of social life, not to be tampered with, or criticised, otherwise this will spoil the rules of the proposed games.

This cuts across the writer's belief and Taylor and Carter's (1969*b*) rejoinder fully sets out the case against such a view. Briefly it is their contention that games may be closed or open to the extent that the designer and/or instructor may require. In some games, for example, MANCHESTER (Blaxall 1965), which simulates the processes of industrialization and urbanization typical of the mid-nineteenth century in the U.K. or Western Europe, the process is inexorable. Play tends towards the situation in which the agricultural improvement reduces rural labour needs, whilst raising the food production level. Migration to the town takes place, increasing the size of the town; the scale of industrial production increases with greater-than-scale increases in output. If the instructor was trying to convey the process of urbanization in nineteenth-century Britain and its causes, such results would appear to be entirely appropriate.

In other circumstances however a more open-ended situation might be required, for example, the potential planner might need to be made aware of the extent to which he, in due course, might be able to influence certain processes to produce a desired outcome. Here, Taylor and Carter's anglicized derivatives of CLUG are representative of such an approach. While the games rely upon a series of rigid processes, how the student interprets his play and influences the occurrence of these processes, and thus their outcome, is for him to decide as an individual or as a team or community member. Behaving in this way, the player follows the same type of procedures as a planner who, knowing the outcome he desires, looks to the processes available, legislative, financial and institutional to bring about the desired outcome.

Certainly, rule and value systems do not have to be frozen when embodied in a game. It seems more likely that, in accord with Benjamin's (1968, p. 525) view, the majority of planning games are, and will continue to be, concerned with means of exposing the student to changing value systems rather than confronting him with a vehicle which supplies finite value judgements. The central concern is with a didactic process which is not a closed one: it is open to change especially from human agents operating within an on-going system. It must be remembered that the structure of the game is what the designer would make of it, and what the operator and players' behaviour defines at an operational level. This behaviour, Schild (1968, pp. 98–9) has suggested, is influenced in part by the time-scales of games sessions. In his experience, shorter game runs tend to emphasize the unethical or the 'sharp practice' as a matter of self-interest. The lengthy sessions go some way to ameliorate this problem by establishing implicit norms or values which are generated over time in response to experience and in response to, or conformity with, wider interests. Alternatively, of course, such norming procedures can be explicitly written into the game as and when operational circumstances demand as in the case of the DUNDEE GAME (Henderson 1968).

Following on from this point, game rule systems can be seen to serve as a useful basis for comparisons with actual real-world rules which are instrumental in generating or shaping urban relationships. Alexander (1965, pp. 168–76) has suggested that these rules are different from game rules in two ways. First, they operate at many geographical scales. For instance, rules which operate at the urban scale control the relationships of streets to one another and rules which operate at the scale of individual buildings prescribe the arrangements of rooms within buildings. Second, the rules in the urban region are enforced by diverse agencies. Some rules are enforced by legal means, as bye-laws or as planning controls; some rules are enforced by inducements and tradition. The significant point, in this context, is that the total number of rules and enforcement agencies operating at any one time is tremendous. Because of this, Alexander has pointed out, it is difficult to think of the totality of rules operating at any given moment, regardless of scale and regardless of the means by which they are enforced. Within a game, however, this situation becomes a comprehensible proposition, rules can be explicitly stated in their relational form and enforcement can be in the hands of a single agency. So the student can be encouraged to think about the total number of real-world rules in a 'systems' context. Thereby the potential planner is given one means of learning to appreciate the value of an explicit rule system and co-ordinated control of that system.

Finally, when considering the questioning of rule and value systems as embodied in games, the fact that this is an extremely sensitive area says something, in itself, for the clarity of the anatomy of the technique. The fact that in games debate can readily focus on central issues to some extent demonstrates the importance of a teaching tool which attempts to spell out its own skeletal form. The point of interest here is not whether a particular value is debatable *per se* but that the debate can rationally take place at all; within an explicit framework a common wavelength is established – without such support it is difficult to know where to start.

It is at this point, however, that it is increasingly evident that, in attempting to present a balanced view of the negative qualities of instructional planning systems, the discussion has come close to setting down the counter argument in terms of positive qualities. So, before such statements predominate out of turn, these issues are now considered.

ATTRIBUTES, BENEFITS AND RELATED ADVANTAGES

If the game model is considered as a component in the instructional simulation system, as in this publication, then, at this stage, the attributes of the fully integrated components or total pedagogic system should be properly enumerated. However, as reviewing the claims advanced for such a conception would be an excessively lengthy process, discussion is restricted to certain of the major assets which have become associated with the technique.

The base from which all gaming procedures emerge is the modelling process and the actual task of designing games is seen by many as being a fundamentally instructive activity which requires detailed data collection, systematic analysis and an ability to recognize and synthesize essential and important relationships. Even when a satisfactory model is not achieved, the investigators are led to believe that they have obtained a considerable insight into the nature of the phenomena under study. For example, the model-building process for the instructor is a means of demonstrating how little he knows about his teaching objectives in terms of the value of each academic task, its associated response and its relationships within the general educational framework. In other words, the formulation of a precise statement of objectives and the level of clear thinking required make it necessary to consider the whole range of the designer's concepts, assumptions and tenets concerning his subject-matter.

In addition to having instructive qualities for teachers, game design obviously presents instructive opportunities for students. Here, games appear to encourage the student's approach to the simulated situation to be more sophisticated and comprehensive. Rather than seeing the idiosyncrasies of a particular situation as in the usual studio project, the student can discover for himself and argue the general validity of the structure of the model underlying the game. If he finds the game model unsatisfactory, then in attempting to improve it, he is questioning his own understandings of the situation modelled, clarifying them and adding to them. Thus, for Cherryholmes (1966, p. 47) redesigning or modifying games is one way of maximizing the usefulness of the technique. Shubik (1963) has further argued the value of this extended approach as a research tool and as a means of enabling the student to move along the continuum from training, through education to research. As a future planner this elementary model-building experience will stand him in good stead, in that it should help him to distinguish, judiciously, between the important and the trivial.

In terms of student appreciation, all advocates of the technique point out that, almost without exception, the operational component of the instructional system is seen, primarily, as an enjoyable experience. All gaming models appear to be able to arouse and sustain a high level of interest and enthusiasm; consequently, the majority of students appear favourably disposed towards the experience and, as has already been said, very willing to take part in lengthy experimental runs outside normal academic hours. Whether a technique which is enjoyable is more conducive to learning than when the reverse is the case is a contentious issue. Suffice it to say that at least one eminent educational specialist (Beard 1967, p. 27) has pointed out:

> . . . it is a well known principle in physics that the less the resistance the greater the effect for the same effort, in other words, the greater the efficiency. This is a principle which holds in learning as well as in physical phenomena: if something which students enjoy doing can be introduced relevantly into the learning the resistance is reduced and efficiency is likely to be maximized.

By way of further comment on this point, it can be noted that the game environment seems to reduce the, so-called, traditional instructor–learner polarization. During operating sessions, no judgement on the students' performance is generally needed since shortcomings are very largely self-evident. The instructor can often stand aside and allow the participant to look upon the effects of his own behaviour. *Team progress* is invariably monitored according to a predetermined scale and so most games are free from a type of subjective teacher assessment which, in some instances, might inhibit involvement or learning. At the same time, the student is presented with opportunities to recognize his own *individual progress* at a level which allows him to derive his own personal punishments and rewards. In other words, the achievement of cognitive skill and intellectual mastery is, in itself, rewarding just as failure is punitive. This is particularly so, according to Bruner (1966, p. 30) when the learner recognizes the cumulative power of learning, that learning one thing permits advancement to something that before was out of reach. To Bruner, and to many others, this is a truth which has been known to every good athletics coach since time immemorial.

Thus, gaming-simulation has come to be viewed as one means of reducing a potential source of disruption which might spring from personal antagonism in teaching situations. It is a departure from the norm which adjusts to a personal level and allows a freer basis for social as well as intellectual exchange. In other words, the game becomes a way of softening the difference between those in authority and those under authority; interstudent relationships tend to predominate as the characteristic procedures greatly facilitate continuous opportunities for student to student learning. Finally, when considering the motivational attractiveness of the technique, it should be recorded that claims in this direction are made with particular vehemence when students have been involved in the model building as well as the operational process. Here the student is seen to receive a fuller insight into the complexities of the system under study and is made painfully aware of the human judgements and areas of uncertainty embedded within the simulation game.

The value of such a receptive learning situation is also thought to be seen at its best when experienced professionals and neophyte students participate together. In such instances, both groups are reported to take part readily, and harmoniously, on an equal footing and reactions from both parties suggest that the benefits, from this type of confrontation, accrue equally to both sides. In this way, games may be seen as a means of combining or drawing together a variety of attitudes and points of view not collectively or selectively revealed by other techniques. Thus, some investigators have used simulations as an educational tool to reduce the so-called ' credibility gap ' between students' attitudes and those of real-world proponents whose actions the students are attempting to portray and understand. By stepping into someone else's shoes, in a statistically structured role playing situation, personal bias can sometimes be ameliorated and ' them–us ' polarities can be replaced by greater respect, increased understanding and even sympathy for a cross-section of real-world views outside the students' normal experience.

Staff from Project Simile (1966) have developed this point and have further suggested that instructional simulation may perhaps act as an information retrieval device by putting together facts, insights and principles already acquired by the student but kept compartmentalized. By emphasizing the interconnectedness of knowledge, gaming-simulation may help to reduce the artificial barriers erected between disciplines and demonstrate the gaps and overlaps between different disciplinary approaches. This may quite well prove to be of significance in fields such as urban and regional studies which are essentially multi- or interdisciplinary. Whilst sets of parallel lectures and seminars, from individual disciplinary specialists, do not necessarily force the student to confront one understanding with another, gaming-simulation, and perhaps studio work, may be claimed to have this facility.

From such a discussion, it follows that games offer particular advantages in the way that they can present an integrated or synoptic view as well as providing a generally freer means of communication. Participants can see the world through eyes other than their own, in the sense that they can occupy roles not normally available to them. The game tends to make all parties involved more explicit about what they see, hear or do and the relative simplicity of the mechanism greatly facilitates the comprehension of a more complete or comprehensive picture. In addition, game procedures tend to dramatize the inter-connectedness of information and ideas and so serve to clarify areas of common concern not readily apparent in more formalized one-sided exchanges. In all these ways, teams and individual team members can confront each other with interdisciplinary expertise and personal knowledge in a manner that seems likely to produce a richer urban dialogue. Undisputed progress in this direction, particularly in terms of the merging of disciplinary boundaries, is already being made by virtue of the composition of many simulation design teams.

One of the activities which is central to much of the foregoing discussion is decision-making in relation to the shaping of man's environment. It will be recalled that, in essence, urban development games are dynamic representations of the human settlement which attempt to describe, simply, the milieu within which the planner works and these abstractions allow certain representative features to be easily understood and manipulated. In short, participants play the game to discover the controlling parameters of the synthetic environment. In entering into this commitment, a number of the facets of the decision-making process are seen to be rewardingly revealed.

First, the decision-making takes place in a risk-free environment. The participant is in no danger to himself, to others, or to expensive equipment and resources. He can afford to make mistakes knowing that his actions do not carry real-life consequences and, as action and feedback are closely related through the compression of time, he can profit from his mistakes by amending his performance in accord with his results. So, on the one hand, the game constitutes a comparatively threat-free environment and, on the other hand, the game aims to demonstrate how experimentation can be both revealing and rewarding.

Second, the manner in which decision-making in games can become sequentially more challenging is regarded by many as an important attribute. The framework of the model often permits the gradual introduction of complexities as student attainment becomes clear, so, in theory, the student can become ' acclimatized ' to a certain situation before being confronted with another level of learning or a task of greater magnitude. This ' acclimatization ' process seems to be achieved at the moment very largely through the addition of ' plug-in ' elements and through the integration of related models. Be this as it may, the undisputed point seems to be that instructional simulation systems can advance learning if judged by their ability to build up decision-making complexity and experience in terms of a logical progression.

Third, and closely related to the build-up of decision-making practice is the discipline of systematically being encouraged to think through carefully staged problems. In this sense, gaming is essentially a process for learning to develop *a feel for* and *an insight into* evolving situations. This abstract quality has been touched upon by Raser (1969, p. 133) who has likened gaming participation to . . .

> the architect, who, before he arrives at a final design, tries to imagine himself *moving around* in his building, who tries to *experience* it rather than simply peering at it from the outside, it may be that the student of social processes gains a more meaningful comprehension of those processes by getting inside them, by experiencing their dynamics in the microcosm of the game, instead of by looking at them from the distance of a book or a lecture.

In short, game decision-making experience is one method of cultivating a well-rounded appreciation of the dynamics of the total situation.

Fourth, one facet of the planner's decision-making training which is often neglected is that of exposure to the irrational. Unfortunately, rationality is not always the hallmark of personal or group decision-making and games often serve to make this explicit. The volatile homo-sapiens can be represented in juxtaposition to more concrete environmental data and the interplay between logic, rationality and irrationality can frequently be dramatized. The strength of the claims on this point rest very much on the ability of the technique to come to terms with the behavioural component *in relation to* more readily quantifiable phenomena.

Fifth, and a particularly valuable aspect of game decision-making is the fact that students, by their actions, can clearly demonstrate their understanding. Simply stated, they are required to prove their worth within a prescribed environment. They are required to *demonstrate* their power to manipulate and manage a simulated system. Each player's general objective should be to exploit the situation as he sees it. Performance is not dependent upon unwritten rules which individuals consider *ought* to represent the system under study; these beliefs may or may not be valid. Instead, interest centres on the display of skill and knowledge exhibited in particular decision-making spheres and through the compression of time, this expertise or competence is rapidly revealed to the individual concerned, to other team members and to the teacher.

Sixth, games also provide a means of incorporating into a curriculum a wide range of decision-making experience. In this respect, few techniques can cover such things as short or long term priorities, local or regional interests, public or private commitments at one and the same time, whilst simultaneously being able to modulate information flows, degrees of uncertainty with regard to participant co-operation, competition, conflict or collusion and amounts of stress as a consequence of time pressures.

Seventh, following on from the foregoing and in conclusion on this point, gaming procedures are seen as potentially very *flexible* frameworks within which to practise decision-making. Not only have they been seen to be able to incorporate different levels of decision-making, diverse varieties of phenomena and open-endedness but they also provide an economical means of examining the evolution of alternative decision chains under varying conditions. This flexibility is claimed to be particularly important in gaining an understanding of causality relevant to the urban process. In other words, the interdependency of actions over time can be repeatedly articulated and contemplated without too much difficulty and without inhibiting free exchange and discussion throughout proceedings.

Finally, when considering the benefits and related advantages of urban development games, a whole spectrum of opportunities exist in relation to the integration of various new planning tools, as decision-making aids, within the learning environment. The richness of these possibilities is a further testimony to the versatility of the technique although, as yet, the majority of claims advanced in this direction appear very much to be indications of promise as opposed to clear indications of proven worth. For example, the growing sophistication which is being developed with regard to the man–machine interface, involving various forms of electronic data processing and presentation, is being incorporated into, and in some cases growing out of, gaming-simulation research. Similarly, on another level, techniques such as discounted cash flow analysis, threshold theory and planning programming budgeting systems are being introduced and refined through rudimentary experimentation within games (for example, see Armstrong 1970). Obviously it is very early to attempt to categorize this ' fall-out ' from gaming or the related ' spill-over ' from an interest in urban modelling but, even at this stage, considerable confidence is being placed on the role of instructional simulation systems both as a catalyst and as a test-bed for new social technologies.

INTERPRETATION CAVEATS

In placing the remarks, so far set out in this chapter, into proper perspective, certain additional observations appear to be required by way of qualification. These caveats are now considered, bearing in mind that several general reservations concerning the survey of claimed advantages and limitations have already been enumerated.

The attitudes, opinions and reactions voiced about instructional gaming-simulations are both wide ranging and far reaching; perhaps this should be expected with a technique which is both new itself and new to a relatively young subject area. In situations where widely divergent views have been expressed on the same gaming model, some explanation can be found in the differing objectives of the users and in the contrasting educational context and the varying mode of operation in question. In other words, considerable discrimination is required before claims for, or against, a specific instructional simulation system become valuable or relevant. To be useful, the objectives of the exercise in question have to be spelt out alongside descriptions of the procedural methods actually adopted and the nature of the specific educational context. In short, more steps to encourage and promote greater precision in planning education are urgently required.

As with any technique, care is needed; gaming-simulations may obviously be employed either efficiently or inefficiently. To achieve satisfactory results requires not only a willingness to integrate them into the educational system but also discerning judgement and a high degree of expertise. The dangers of erroneous transfer and abuse can to some extent be mitigated by careful model construction and skilful operational control. In its ability to teach the ' right ' rather than the ' wrong ' things, gaming appears to be no more foolproof than any other technique.

A number of cautions have been voiced about the emotional problems which may occur in a gaming situation. For example, initially some participants find it difficult to accept a simple, yet highly formalized game as a substitute for a system of complex, environmental relationships. This scepticism is understandable when viewed against the novelty of the technique and the educational background of the players. It is not too much to hope that this attitude may be modified, over time, as a freer and more experimental approach to education becomes more commonplace. In addition, the widespread usage of academic games in the school system plus advances in the technique at a professional level will bring publicity and familiarity which *may* dispel unjustifiable inhibitions. It is then likely that a greater problem will arise from conflict between games as entertaining devices and as educational instruments; already, there are those, in other fields, who believe that games produce more enjoyment than learning (Kraft 1967). However, it should be recalled that planning education has always had, via ' studio ' work, open and informal simulation teaching situations and the current level of staff and student support may well be a reflection of past experience. Consequently, this may indicate that planners' initiation and acceptance problems with games might be fewer than in some other areas lacking a similar tradition.

One final qualification follows this brief survey of claimed advantages and limitations. Such a collection of assertions is built very largely on a growing consensus of opinion rather than on a well-documented catalogue of proven successes and failures. Since learning situations differ one from another, what is a satisfactory combination of human and technical resources in one case may not necessarily be an appropriate ' mix ' in another. Much rests upon the quality of the game itself and the skill of the administrator in directing all efforts toward the assigned objectives, It must be acknowledged that few validation studies on gaming techniques are available to shed any real light on the efficacy of *particular* simulation procedures in *specific* educational settings. However, this is a predicament seemingly common to both new and old techniques in higher education.

Thus, the case for or against this form of simulation as a learning technique stands very largely unproven. Having said this, one must bear in mind that all forms of instructional simulation are relatively new and planning games, in particular, constitute a very recent development in this field. In addition, it must be remembered that, although a considerable amount has been published on the educational value of gaming procedures, few dissenting voices can be traced in the literature. Whilst critical reports are apparently absent, the literature is regrettably characterized by two other features. The loose, often extravagant claims made for the value of the technique is one common feature; the other is the scant nature of descriptions of particular applications of the technique. In sum, this is a very new field where there is a scarcity of sophisticated evaluation in comparison with the richness of subjective reports, often pitched at an anecdotal level.

In summarizing this chapter, certain points bear repetition. First, planning gaming is increasingly becoming a popular preoccupation. Second, as yet the technique cannot claim to be fully integrated within the curriculum of many urban studies programmes and, so far, it has only occupied a minute position in the student's timetable. Third, the preparation time and effort behind such a small component of the timetable appears formidable. Few have challenged this criticism and further experimentation and development is being continued on the assumption that the time and effort required may well decrease as familiarity and expertise increase. Such confidence is very largely based upon experience in other fields and the increasing appearance of commercially produced ' packaged ' materials. Set against this, instructional simulation systems offer special levels of simplicity, involvement and experience in coping with complex and highly dynamic decision-making processes which can be manipulated as well as widely understood and these properties represent, for many, the most convincing arguments in support of the activity.

The one dominant feature of the literature and the writer's 1969 survey is one of widespread enthusiasm and considerable faith in gaming-simulation procedures. This enthusiasm appears to be undaunted by the scarcity of validatory material or by the modest place the technique often holds in many higher education timetables.

Finally, it is important to accept that games, in addition to their fundamental advantages, have certain attributes which apply equally to other teaching methods. However, an effort has been made to emphasize those qualities which appear particular to urban gaming-simulation. In addition, the technique has certain shortcomings which may, or may not be common to other teaching tools. The positive and negative properties presented thus far do not, of course, validate or discredit the method as a whole. An overview has hopefully been established which deepens the reader's understanding and, more specifically, increases his awareness of possible assets upon which to capitalize or potential drawbacks to ameliorate.

6. A gaming-simulation rationale

In the previous chapter, after setting down a picture of instructional simulation usage, a large number of claims were advanced regarding the properties of gaming-simulation procedures used in the study of the urban development process. At the same time, only two validatory studies were discussed in terms of the efficacy of the technique and consequently it is felt that some clarification of the rationale behind the use of planning games is now required.

At the present state of the art and particularly bearing in mind the scarcity of empirical data it is contended that it would be presumptuous to attempt to outline a tight and formalized theory supporting work in this field. However, it does not seem in any way premature to take stock of some of the assumptions which have, or appear to have, shaped developments to date and as far as possible clarify and order these assumptions. This chapter is thus an attempt to put together an account of the influence of several movements and individual initiatives which have contributed towards the building of a body of theory relevant to urban gaming-simulation. It supplies several perspectives on the evolution of the technique and culminates with an identification of the emerging logic which might be said to structure and support much of today's instructional simulation concerned with the urban process.

Kilbridge (1968, p. 382) has reminded us that one measure of the development of a field of knowledge is the extent of its structured theoretical base. He has pointed out that such a base is constructed from the accumulation of many theories, narrow and broad, specific and general, overlapping, contributing and conflicting, until enough have accumulated to be articulated into a structured base. In these terms urban gaming-simulation 'theory' is seriously underdeveloped, ill defined if not non-existent. In its place and in the current climate of conjecture it may be worth-while to draw from empirical studies, separated in time and place, such generalizations as seem consistent with the need to provide, however tentatively, an explanation as to why gaming-simulation procedures seem to be particularly attractive to those studying the urban process. In other words, is it possible to establish what prompted planners to adopt and continue to use the technique over and above the attributes already described? Given that the complete overall theory of gaming-simulation for pedagogic purposes seems to be a long way off, what premises underlie continuing efforts in this field?

To provide answers to such questions it is proposed to draw upon diverse theoretical support from various fields which may not necessarily interlock and yet seem to provide a base from which greater understanding and further development and research can be advanced. The interest, here, is centred on the elucidation of tentative foundations from which to consider further action. Support is sought from the collective endeavours of many who have been concerned, in the main, with learning, systems analysis and planning; this means that the rest of this chapter is devoted to research which might shed light upon, on the one hand, aspects of intellectual development and the learning process, and on the other hand, upon efforts to achieve greater understanding of the urban process and the management of change. Bridging these two fields and permeating both areas are several new social technologies and these new ways of viewing complex systems will be seen as the common link between learning, planning and instructional simulation systems of the urban development process. Some of the principal contributions in each of these areas are now examined after first setting out to summarize, briefly, the totality of interest which has long been centred upon games as a universal behavioural mode.

GENERAL HISTORICAL PERSPECTIVE

Throughout history, various forms of games have been the basis for a plentiful and contentious body of literature. Gaming procedures have attracted a profusion of scholars who have sought to derive, amongst other things, various psycho-analytical insights, social and ecological understandings, mathematical as well as economic theories, educational principles and even philosophical hypotheses after studying games. Such works as: Berne's (1964) *Games People Play*; Caillois' (1961) *Man, Play and Games*; Piaget's (1962) *Play, Dreams, and Imitation in Childhood*; and Von Neuman and Morgenstern's (1944) *Theory of Games and Economic Behavior*; bear witness to the complexity of the subject and the diversity of scholarly research stemming from an interest in gaming procedures. Unfortunately for this discussion, the results of so much and such diverse study appear to have divided the subject until the unity of the activities studied in the name of ' games ' is no longer readily discernible.

Because of this richness in the literature, there is always a danger that any selective presentation of material errs from the impartial. However, as the present concern is one of establishing a rationale for simulation games, the identification of certain positive characteristics would seem to be in order. But before examining the relationship between gaming-simulation procedures and certain ' concepts of learning ' it seems necessary to remind the reader of some of the general, if complex, educational concern attached to games and so-called ' play ' activities.

For example, Dewey has discussed games in his various works to such an extent that it is difficult to summarize even one man's findings in a review of this nature. However, given the present constraints, certain findings do seem particularly significant: first, at a general level, Dewey was an active supporter of creating in the classroom real life situations which allow the active and simultaneous participation of both students and instructors; second and more specifically, Dewey viewed games as an integral part of the learning process and not simply as relief from the tedium and strain of *regular* work; third and finally, it is clear that Dewey thought certain varieties of games could have a definite moral value.

Similarly, other school system pioneers have also recognized that the playing of games deserved a special place in teaching. Montessori (1909, 1914 and 1965), Pestalozzi (1819) and Piaget (1962 and 1965), amongst others, have examined and advanced personal theories on the usefulness and validity of the activity. In this context, Piaget's conception of games as an introduction to life seems to be particularly relevant. Here the game, for Piaget, has been seen to provide an introduction to real-world rules and changing constraints on behaviour, an introduction to group working and the problems of inter-personal relationships, and an introduction to the idea of working towards individual and collective goals involving varying degrees of co-operation, competition and conflict. In passing, it is interesting to note that this link with life has been recognized more recently at a psychiatric level in the work of Berne (1964) whose research has been popularized by his book *Games People Play*.

This approach to games has also been adopted by Coleman (1965, 1968*a*, 1968*b*) who in answering the question ' What fascinates me, and others in my field, in the art of constructing games? ' has seen gaming procedures as a ' kind of caricature of social life ' – an abstraction which allows fundamental learning from the observation of a gross simplification. More importantly, perhaps, he has also seen games as vehicles for simultaneous learning at all levels – fascinating to both students and scholars at one and the same time (Coleman 1965, p. 4).

Obviously the fascination of games is an elusive quality which few have attempted to rigorously define. The literature on ' games ' and ' play ' is equally contentious and plentiful and for the most part so intertwined that distinctions are difficult to make and embarrassing to substantiate. Not all ' theories ' of play appear to fit all varieties of play and many of the attributes of this behavioural mode are equally present in other activities as well. The majority of assertions concerning play as an educational activity have their foundations in primary education and it remains to be seen how valid these concepts might be at later stages in the educational process. Despite these caveats, the fascination of ' play ' and ' games ' remains very largely unchallenged and the attractiveness of the behavioural mode cannot be overlooked.

Indeed, if anything can be said to emerge clearly from a survey of the literature, it is the fact that games undoubtedly have an ' *association value* ' for the majority of students. In short, they are pleasurable activities which possess various degrees of commitment in terms of voluntary participation and challenge at both group and individual level. Having said this, the burden of much of the literature is devoted to the propagation of the wider relevance of games outside the frequently accepted realm of pure fun or simple entertainment. In summarizing this situation Inbar (1969, p. 29) has, quite rightly, reminded us that we know practically nothing about games in a general sense other than that some are interesting and others are not, and in all probability children appear to learn from *some* of them. Such a modest conclusion seems to fairly indicate the present state of knowledge relating to this universal form of behaviour which is increasingly being seen to have potential within the learning process.

CONCEPTS OF LEARNING

From the preceding general historical background it is not too difficult to go on and identify in broad terms certain principles of learning which are acknowledged and followed by many instructional simulation systems involving gaming procedures. These generalized ' laws of learning ' have tailored, in many ways, the form and operation of simulation games although, to date, few designers or users have felt free to explicitly acknowledge their pursuit or recognition of these conditions. However, there are at least three exceptions which appear worthy of comment because their work appears to attempt to relate, however tentatively, instructional simulation systems to what is known about intellectual development and the learning process. These contributions rely heavily upon first hand, personal interpretation and sectarian interests but, bearing in mind the scarcity of such material and the weight of each author's total experience, their pronouncements are now considered.

The conditions favourable to learning as defined by psychology have been set out by Greenlaw *et al.* (1962, pp. 32–49) with specific reference to the deliberate embodiment of these conditions in business simulation exercises. These six ' laws of learning ' which have been to some extent respected by gaming-simulation procedures can be summarized under the following Greenlaw *et al.* headings:

(a) *Contiguity*

The proximity of ideas or events plays a large part in the learning process, hence in any instructional simulation. Effective learning results from the clarification of cause and effect relationships; in a game, actions have direct consequences and the instantaneous feedback on the adequacy of performance provides a prompt means of reinforcing appropriate actions.

(b) *Effect*

It is widely held that the more satisfying the result, the better the chances of learning. Games tend to score highly here with their undisputed power to motivate and involve students over long periods and even outside normal academic hours. The theory, in other words, considers adequate motivation a desirable prerequisite to achievement of the required behaviour.

(c) *Intensity*

Following on from this point, it has long been recognized that learning increases in proportion to the degree of total student involvement. As gaming procedures are not only highly motivating but demanding in terms of their call upon the full range of human senses, it is deduced that their effectiveness is likely to be greatly facilitated.

(d) *Organization*

Learning sequences structure the speed of learning and one of the strongest attributes of most games is their clear and orderly structure. To maximize the learning experience resulting from this clarity and order, further levels of organizational skill are brought into play. In essence, the game supplies the fundamental organizational framework upon which the administrator, the player, the team and, in some cases, the umpires, can build to promote increased achievement.

(e) *Facilitation and interference*

Students are also widely encouraged to build upon previous experience so that moving from the 'known' to the 'unknown' can occur in a way which will facilitate this advancement. The removal of background noise or irrelevant interference can obviously assist the learning process. Games recognize these principles by abstracting the essentials from a situation and, on occasions, assist the transition from the 'known' to the 'unknown' by progressively increasing the complexity of simulation tasks as play progresses.

(f) *Exercise*

Enduring learning requires practice ('doing'), hence the adage 'practice makes perfect'. The embodiment of this final Greenlaw *et al.* principle is very much self evident in instructional simulation systems as they are centred, almost without exception, on sequential decision-making involving continuous and repeated learning cycles.

An alternative view of learning conditions relevant to instructional simulation systems has been set down by Greenwald (1966, pp. 20–1). After considerable experience in using business games as a teaching device and, after becoming familiar with the related literature of educational psychologists, he has set out ten concepts of learning which correlate the views of some of the leading theorists and, at the same time, support the use of instructional simulation. These ten concepts are defined concisely by Greenwald (1966, pp. 20–1) as follows:

(1) Learning is re-enforced by *Repetition.*
(2) Learning is more effective if more faculties are brought into the learning process (sight, hearing, writing, *Thinking*).
(3) Learning is enhanced if the dissatisfaction with the status quo comes from within (*Desire* in the trainees).
(4) Learning is more effective where the student plays an *Active* part in the learning process.
(5) Learning is more effective in a *Small Group* working towards specified objectives.
(6) Learning is enhanced by personal *Involvement.*
(7) Learning is more effective if the situation has (at least) the appearance of *Reality.*
(8) Learning is re-enforced by a prompt *Feed-back* of the results of previous exercises.
(9) Learning is enhanced by a *Contingency* between present and past learning.
(10) Learning is enhanced if further material brought into the learning situation is in the *Same* problem areas.

Finally, Duke, after viewing the various efforts to evaluate gaming procedures in conjunction with his own extensive experience in designing, developing and testing planning games, has suggested a series of criteria which must be met by an effective gaming instrument. Duke (1964, pp. 37–8) has seen these criteria as being ‘ universal ’ requirements for an effective game regardless of subject-matter and has set down these requirements as follows:

(a) *The prime purpose of gaming-simulation techniques is to provide an environment for self-instruction.* The technique is particularly well suited for conveying concepts which have not been made explicit in theory.

(b) *Gaming techniques are particularly valuable for conveying concepts of elaborate systems* through the device of employing a simplified model. This provides the student with a rudimentary framework useful for the retention and categorization of associated learning experiences, for integration of diverse materials, and as a platform from which new experiences may be evaluated with a resultant expansion of the complexity of the internalized model.

(c) *Emphasis must be placed on enhancing learning which is general and structural.* The technique must be effective in integrating decision processes as well as specifics and in bringing into meaningful focus material derived from conventional educational experience.

(d) *The technique must be integrated with conventional teaching experiences.* This is necessary to ensure an equitable distribution of teaching resources; further, to obtain maximum benefit from the experience, there must be a careful grading of the complexity level of the game to the students' level of ability and curriculum content, insuring mutual reinforcement of learning.

(e) *The game will force player behaviour patterns to be explicit,* in terms of: defining the player's own role, the various roles of other players, and the interactions among them; increasing awareness of the complexities resulting from these interactions and of the dynamic nature of the interaction; forcing a summary evaluation by the player of a new and uncertain environment and the consequent adoption of his decision framework; problem definition, goal formation, the development of long-range plans and alternatives for implementing these, and policy definition.

(f) *The game will enable specific skill development*: by providing for the introduction of specific techniques; through the presentation of factual information in appropriate context; and, which will have a high correlation with skills required for competence in the profession.

(g) *Effective gaming requires a vehicle of sufficient realism*: to generate enthusiasm (as in the case with intra-mural sports) to ensure voluntary participation; to be convincing in the relationships and specifics which it portrays; to create an intensity of group interaction of a sufficiently high level to be of value in building stress immunity in the players; to prevent erratic or unrealistic player response relative to that expected in the ' real world ' constraints; and to be capable of creating the high level of motivation required for learning.

(h) *Finally, gaming requires a mechanical vehicle*: which is reasonably convenient to operate; which enables modification and referee intervention when appropriate (' free ' game); in which design variables are not significantly regarded by random (non-design) happenings; and in which feedback mechanisms are both quick enough in response and sufficiently intricate to influence player behaviour.

The preceding three check-lists have much in common and reflect upon the unsophisticated level of early efforts to formulate a theory to support the use of gaming procedures in instructional settings. It is true that the Ph.D. dissertations of Greenwald and Duke expand upon specific validation problems in this area but they claim little in advancing the reader's basic knowledge of a generalized instructional simulation theory or understanding of a multi-faceted approach to validation after setting down the broad concepts of learning which have just been quoted. To add to these statements, it seems necessary to draw upon certain selected learning ' theories ' and this is now attempted.

For example, Beard (1970, p. 6) has pointed out that gaming-simulation procedures come close to offering all things to all men in as far as they combine some of the features recommended by two prominent ' schools ' of learning. In so far as simulation systems encourage the student to follow a set activity and indicate the quality of performance at sequential intervals, they resemble linear programming techniques developed by the psychologists of the ' associationist ' school. The associationists tend *not* to favour learning concepts based on the introspective qualities of the learner and, in preference, view learning as a process built up by reinforcement to responses triggered by certain stimuli. Thus, they support an approach which relies upon specific responses meeting with ' success ' and hence learning, whereas ' unsuccessful ' actions are unrewarded and the learner is encouraged to seek a more appropriate ' programmed ' response. Following this reasoning most gaming procedures are broken down into small steps, common in programmed texts, so that the consequences of particular responses can become evident.

On the other hand, Beard has recognized that, despite their careful organization, instructional simulation systems may also allow a degree of choice and some originality into the pedagogic process. In this respect they support the learning approach advocated by ' field ' psychologists. The so-called ' field ' psychologists tend to describe learning in terms of individual interpretation of problem situations in a way that is meaningful to the individual's background and knowledge. Accordingly, they are often concerned with ' open-ended ' learning experiences which are seen to allow the learner to develop his own preferences and to follow his own goal. Hence certain games appear to favour this reasoning when they offer relatively unstructured opportunities for the participant to gain ' insight ' from ' self ' interpretation of situations and from personal interpretation and use of information.

Thus, Beard (1970, p. 6) has seen instructional simulation systems as being capable of combining some of the approaches recommended by each of the two mentioned ' schools ' and has suggested that this might explain why such activities are almost universally enjoyed. In particular, the weaker students are seen to be able to learn at their own speed, yet within a supporting framework, whilst the more able students are not too restricted to try out more advanced exploration of decision alternatives, often at one and the same time.

Amongst those adding to a wider appreciation of simulation games the influence of Skinner has been significant; indeed in his efforts to develop a technology of learning his work is likely to be one of the major landmarks in the evolution of educational practice. Skinner (1953, 1963) holds that behaviour is controlled by its consequences and anything which is to be learnt must be divided into a large number of small successive steps and that the student's accomplishment of each of these must be positively reinforced. In other words, the learner must be assured that he has grasped each step and has made the appropriate response before proceeding to the next task.

In sequential decision-making procedures, such as are embodied in the majority of simulation games, some of Skinner's principles are respected alongside several other important conditions for learning. For example, the decision format of simulation exercises demands specific responses from the student who must, of necessity, engage in some form of 'commitment' with the learning process. Following on from this commitment, if the student is a member of a team he is made to appreciate some of the alternatives open to him as identified by himself as well as by his colleagues. Finally, as the consequences of his actions and those of his associates are often immediately apparent, he obtains direct feedback on the quality of his decision-making before proceeding to another step in the incremental learning process.

A final elaboration of the concepts of learning relevant to instructional simulation systems has been set out by McLeish (1970). Here, McLeish has sought to define and show the inter-relationship between a system, a model, a simulation and a game, with special reference to education. The principles underlying games and simulations are stated in terms of their particular attributes and in relation to a general 'systems' perspective. As McLeish (1970, p. 11) has pointed out, the path from systems theory to gaming procedures in education is a relatively short one; a path in which the idea of a model is an important stepping stone. In short, the model, the simulation and the game are all tools for grappling with complexity in an attempt to identify particular structures and processes in a piecemeal, yet clear, logical and systematic manner, as far as possible freed from background noise or irrelevance. As this perspective is fundamental to one of the important, emerging conceptions of planning, such views are now considered more fully with respect to the urban development process.

A 'SYSTEMS' CONTEXT

Throughout this work, in passing, reference has been made to new ways of viewing complex systems. In chapter one's contextual outline, these new ways were seen to have freed planners from a mechanistic, deterministic view of the city and region and to have provided the profession with more dynamic and comprehensive tools for conceptualizing man's settlements (both habitat and society) and for influencing their process of change. This, in turn, has been accompanied by new conceptions of planning education and chapter two went on to set out various dynamic or simulation approaches to the learning process which might be relevant here. Now, it is time to retrace these earlier steps in an attempt to inter-relate, more clearly, ‘ systems analysis ’ concepts and the new social technologies with planning and with the evolution of instructional simulation systems. Thus, the rest of this chapter will be concerned with issues which prompted urban planners to adopt and develop innovative pedagogic systems and will seek to identify the justification which appears to support their continued use for instructional purposes.

One issue central to the whole of the ensuing discussion is the emergence and expansion of ‘ systems ’ thinking. This revolutionary approach to rigorous enquiry is of critical importance to this work although it should be pointed out that it is by no means solely applicable either to education or planning or even to simulation games in particular. In essence, a ‘ systems ’ view provides an all-embracing approach to problem situations where a system is conceived as a complex whole – a set of interacting component parts seen as an organized totality. Thus, for example, a system can be real or conceptual; a system may include components which constitute subsystems; a system may be open or closed or alternatively mechanistic or adaptive. Examples of systems and of their classification are legion and, at this stage, it must suffice to point out that the core of ‘ systems analysis ’ thinking is in fact a scientific basis for establishing a synoptic understanding of complex situations. It is a notion which has rapidly gained ground since the advent of World War II and owes much to a growth in interest in general ‘ systems ’ theory, operational research, cybernetics, management and computer science as well as the more recent wider awareness of the relationship between transportation engineering and land use studies. Each of these technologies has made its own particular contribution both to the study of planning and to the development of gaming-simulation procedures. This contribution in relation to the planner’s commitment to the urban process has been well documented by Chadwick (1966) and McLoughlin (1967 and 1969) and fully outlined in connection with instructional simulation by Raser (1969).

Consequently, it is proposed to leave the ‘ systems analysis ’ revolution very largely as a self-evident truth and, in preference, turn to an examination of the focus of much systems thinking which rests upon modelling or simulation. Models in planning have yet to receive such comprehensive and definitive treatment as ‘ models in geography ’ (Chorley and Haggett eds. 1967) and yet the planner’s relationship with models is daily becoming more clear. In simple terms, when a planner, or any decision-maker for that matter, is called upon to prepare a plan or formulate a policy statement, he bases his judgements on the maximum amount of data and skills he can command. Faced with complexity and a wide range of multi-disciplinary expertise, the model can be viewed as a mechanism for extending man’s perception of problems and as an organizational framework for ordering disparate, specialized insights. As planners are concerned with the evolution of an ongoing process and with their ability to anticipate and manage change, the variety of model perhaps of most value to the planner is the dynamic model or simulation.

For most phenomena there are many possible representations or types of model; the appropriate model very largely depends upon the nature of the phenomena *and* upon the question to be answered. Quade (1964, p. 66) has pointed out, for example, that a town can be modelled by a map if the question being asked is how to walk from point *A* to point *B*; but if the question being asked is how to speed up the traffic flow between the same two points a much more elaborate model will be required. Unfortunately, there are, as yet, no ' universal ' models -- that is to say that, at the present time, no one model appears to be able to answer all questions about a given activity. Thus, before selecting a modelling approach, the planner, or more specifically in this context the planning instructor, is well advised to reach certain conclusions about the nature of the phenomena he is proposing to handle and, at the same time, have a clear idea about his pedagogic objectives.

On these points the work of Duke and Meier has been particularly illuminating, especially with regard to the utility of gaming-simulation procedures in planning, and the ensuing discussion will very much rely upon their contributions in this area. First, on the nature of the phenomena handled by planners, it is being increasingly recognised that the evolution of the urban development process is an extraordinarily complex and highly dynamic activity. In simple terms, it involves both *physical* and *social* systems; here lies the heart of the problem, namely the simultaneous handling of *both* ' *types* ' of system as they evolve and interact. On the one hand the physical system is relatively simple to measure and represent as tangible elements are involved. The components of the social system, on the other hand, are not so convenient to handle, as volatile human behaviour is very much involved.

As Rosenhead (1968, p. 289) has pointed out, the social sciences have suffered perhaps by the systems analysts' commitment to industrial, financial or military situations in which the human element could be largely disregarded or avoided. Be that as it may, the position now appears to be slowly changing and now both the management and social sciences are increasingly considering experimental approaches to the identification, quantification and representation of human factors. One product of this interest has been the expansion of gaming-simulation developments. There is every indication of a slowly growing body of empirical knowledge covering human behaviour and societal functioning which is likely to aid the transference of the growing ' systems ' expertise from the physical to the social field and every indication that the limited success achieved in this area in the past in no way inhibits further progress. However, it seems fair to point out, as Raser (1969, p. 23) has done, that current social simulations appear to be relatively unsophisticated and tend, in comparison with simulation of physical systems, to appear rather primitive.

The lack of adequate tools to deal with problems of infinite complexity and richness has created the need for compromise and a partial acceptance of less precise instruments. Here, Ray and Duke (1967, p. 5) have indicated that when complete and detailed theories of process and structural linkages are lacking then partial representations, as well as ' best guess ' approximations, become the building blocks for the construction of tighter, more rigorous theory. The line between rash observation and considered postulation is difficult to draw in such discussions as these two authors freely acknowledge. Thus, for Ray and Duke, gaming-simulations are seen as but *one* way of linking such diverse areas as social psychological theories of decision processes and attitude formation, theories of optimal decision-making, and models of large-scale societal structures, so that research on each front benefits from individual area advances. In short, simulation games appear to be *one* means of establishing an intelligent synthesis of many techniques and a given level of knowledge at the present point in time.

Implicit in this transference of ' systems ' expertise from the physical to the social sciences, in order to come to grips with what for ease of reference we might term the *qualitative* as opposed to the *quantitative*, is the belief that social behaviour can be systematized. Most social scientists readily agree that human behaviour is fickle, volatile or irrational, yet there is no body of agreement on whether these characteristics are subject to measurable determining factors. However, planners using gaming procedures have, by their interest and actions, declared at least a passing commitment to a belief that it *may* be possible to come to terms with what Abt (1968*b*, p. 25) has described as the muddy variables of attitudes, feelings and behaviour and their interaction with the clearer variables of economics and technology.

A second major point on the nature of the phenomena handled by the planner follows on from this discussion and concerns not only the diversity of data under review but also the complexity of planning situations where frequently the extent of the data requirement relevant to decision-making is often unclear. The more complex the situation the more fallible becomes the planner's ability to manipulate and abstract the relevant. Hence the adoption of simulation techniques as one means of extending the planner's awareness of complex and dynamic situations. The advent of the electronic computer and associated data banks is not an answer in itself and one of simulation's great attributes lies in its capacity to simplify through a spectrum of precision ranging from elementary role-playing exercises to elaborate man–machine models.

Here, gaming-simulation procedures can be seen as filling an obvious gap between the generalized, verbal model and the complete mathematical model.

On the one hand, the spoken word is not an ideal vehicle for models because of the multiple meanings of many words and the tortuous problems associated with the manipulation of verbal statements. Numerical representations or mathematical models, on the other hand, are far tighter, more precise vehicles which have, as yet, restricted coverage in the social sciences. Thus, it is important to appreciate that simulation games have certain properties not combined in other techniques. Not all the elements of the urban system can be readily expressed in mathematical terms and the involvement of people, as individuals and socially through institutional, administrative and political processes means that a human non-rational element is a crucial part of the system. Thus, gaming-simulation with its mixed strategy approach to the coverage of the quantitative, through numerical or iconic representation, and the qualitative intangibles, through human representation, appears to be an appropriate compromise. In addition, this hybrid form of simulation retains the basic properties of simulation systems with respect to their ability to collapse both space and time to a degree of abstraction which is meaningful to many academic levels.

Meier and Duke (1966, p. 5) have been more explicit in summarizing the preceding points and have formulated, for planners, a series of propositions with regard to the style of simulation appropriate to particular circumstances. When organizational needs allow a choice to be made between different modelling approaches then Meier and Duke have seen the major sources of information affecting planning policies as a determinant of simulation style. Consequently, they have set out (1966, pp. 5–6) a range of simulation techniques and the context into which they best fit as follows:

If: the basic data are produced by natural phenomena or a large number of small, independent transactions and the final design must produce a balanced network . . .

Then: sampling procedure can be developed which fits into an outline description of the total system (that is a model of the system) that can be manipulated with the aid of a computer . . .

If: the problem seems to be that of acquiring insights into organizational behaviour under conditions in which resources are truly scarce, so that competition becomes intense, and the essential data are qualitative or subjective . . .

Then: a gaming procedure can be developed which allows surrogates of the chief competitors to test the principal strategies open to them and so discover what new and unexpected situations may arise . . .

If: the planning context includes budgeting, marketing or some other well-formed procedures for routinized decision-making, and a variety of poorly defined situations . . .

Then: the simulation should employ a mixed strategy that allows the combination of a simple computer simulation with a series of gaming exercises . . .

So much for the phenomena handled by the planner and his consideration of the urban process in a systems context. To turn now to two general remarks which help to place instructional simulation into a wider perspective with regard to both education and planning.

First and in addition to the parallel with other urban modelling activity, in response to the need for better means of grappling with highly dynamic issues of growing complexity, gaming-simulation appears to fall within a much wider 'frontier' movement which might be considered to be changing the planner's perception of his professional problems. Such a perspective has been provided by Bayliss (1968, p. 3) who, for discussion purposes, has viewed environment as an equilateral triangle with social, economic and physical angles. He has seen a changing emphasis placed on these three facets of environment commencing with a shift from physical to economic orientations occurring over the past two decades and now an indication of a movement towards greater recognition of the social aspects of planning as

Figure 7. *The changing emphasis in environmental planning*

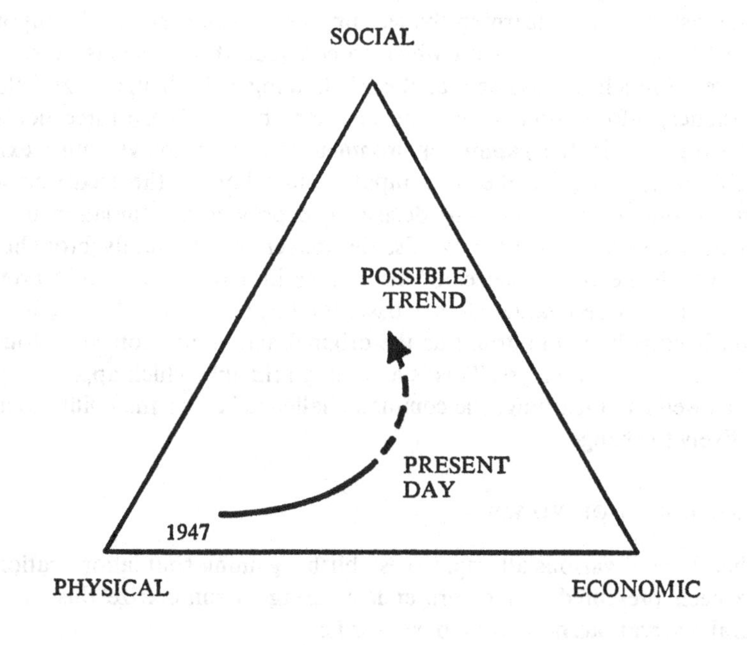

Source: Bayliss (1968, p. 4).

illustrated in Figure 7. This shifting emphasis is well reflected in the development of instructional simulation systems where all three facets of the environmental paradigm are brought together in the planning game in an attempt to present a more balanced, well-rounded didactic view.

Second, gaming-simulation is a part of, and in some cases represents, a new and liberal approach to education. This relationship and symbolic significance is difficult to pinpoint but may, in part, be attributed to: the definition of educational objectives demanding differing methods of communication; the lack of conviction in traditional teaching methods; the reaction to doctrinal enthusiasms and the conventional wisdom in education; the paucity of innovatory instructional techniques for stimulating new interests and opening up new fields of study; and finally, the apparent shortage of ways of cutting across narrow disciplinary teaching. In short, the continued growth of 'activity methods' in the school system has been paralleled in higher education by the increasing use of project, problem solving and seminar techniques as an alternative to extensive reading assignments and the more formal lecture. The emerging emphasis in higher education is one which requires the student to: think for himself; become involved in the teaching situation; find out very largely independent of the teacher; and, above all, *learn to learn* as opposed to becoming learn*ed* – all properties characteristic of gaming-simulation procedures.

The last remark concerning the learning process is particularly important and a fitting observation with which to conclude this 'systems' discussion because so much of 'systems analysis' thinking is built upon the fields of cybernetics, information theory and decision theory. These three fields are drawn together in the gaming environment where the player must explore his situation, recognize the data inputs, and adapt to the feedback loops from the outcomes of previous decisions to provide a foundation for subsequent decisions. In other words, the player is continually brought face to face with the consequences of his actions in a way that allows previous performance to be evaluated as a basis for future actions. Thus, corrective action is equally as important as the original action and control is founded on the correction of errors. This is a guiding principle which applies in planning as well as in learning; the common challenge being: the ability to adapt effectively to change.

A SUMMATION OF VIEWS

In this chapter, various attempts to establish a gaming-simulation 'rationale' have been presented and before endeavouring to summarize this material, several reservations now seem to be in order.

The obvious danger in reviewing such a rich collection of material is that one draws upon only those references which suit one's case. The intention here has been to introduce the nature of some of the research relevant to particular instructional simulation systems and at the same time endeavour to highlight the emerging logic structuring and supporting the use of the technique in a specific area. If anything can be said after such a brief exposition then it must be preceded by such a caution.

Similarly, it must be accepted that much of the material selected is the subject of considerable conjecture and many implicit assumptions, held by workers in this field, have yet to be made explicit. Much of the current interest in, and vigour attached to, gaming-simulation procedures obviously stems from the wider interest centred on what has been termed the new social technologies. Hence, it is often difficult to unravel the interwoven strands of theory which extend beyond one area and provide the logic behind more broadly based activities. A completely satisfactory statement therefore is not to be expected but, in its stead, a number of principles, and the theory which adds weight to them, do appear relevant and should command wider attention.

Following on from this statement, it seems necessary to point out that the word ' theory ' has been used often rather generously throughout this chapter, especially when it is recollected that it is a term rather like the label ' gaming ' which means different things to different people. O'Connor (1957, p. 110) has even gone so far as to suggest that the word ' theory ' as it is used in educational contexts is generally a courtesy title! Here, in place of well argued and established theory a sense of an emerging theoretical base has hopefully been created without resort to tight and circumscribed definition.

In passing, it should be noted that the absence of a theory of instruction as a guide to pedagogy has long been universally recognized. In particular, Bruner (1966, p. 31) has noted that in place of an integrating, educational theory there is principally nothing more than a body of maxims. ' Not only is there no theory of teaching ', Beard considers (1967, p. 1), ' but theories of learning are too numerous and too little concerned with human learning to provide a framework for action.'

Perhaps it is misleading to expect to produce a body of theory equal in coherence and applicability to that encountered in the physical sciences. Certainly, one has to accept that as one is considering people as opposed to inanimate objects, the application of any theory must take account of the dynamic complexity of learning situations. Such situations involve intricate relationships among students, between students and instructors and between the total educational environment both inside and outside institutional settings.

In the light of the preceding remarks, it is difficult to draw together the claims thus far advanced as so much in this new field remains to be substantiated. What emerges from the literature, for the writer, is an increasing awareness of the instructional potential of games for a great number of situations and academic levels. The acceptance of games as a fascinating behavioural mode, to be exploited within the learning process has noticeably so far been achieved almost without challenge. It may be that the word ' game ' is too general to be useful but despite this semantic problem there is little doubt that the many disparate forms of behaviour considered under this heading involve a very popular range of human activities and interests. Thus, planners have become engaged in the cultivation of an age-old behavioural mode which has sought to establish a fuller appreciation for gaming procedures and their value within a total system of instruction.

By way of elaboration, at this point in time, it is only possible to conclude this exposition by acknowledging that, in terms of the principles of learning, instructional simulation systems appear to take cognisance of *at least* four seemingly important educational trends which value:

(a) *active and extensive student involvement* in the learning process, with considerable emphasis placed on guided self-discovery and opportunities to give as well as to receive;
(b) *decision-making experience* in realistic settings with rapid and repeated feedback; the feedback indicating the consequences of actions and the adequacy of performance;
(c) *conducive environmental conditions* with, for example, opportunities to experiment with little at stake and situations where the role of teacher as critic or judge is partly suppressed by self-pacing and self-monitoring instructional procedures;
(d) *diversity of presentation* through differing combinations of media with methods and materials which call upon the full range of auditory, manipulatory, verbal and visual skills.

Similarly, in surveying planning education, it is possible to view planning games within a spectrum of techniques which stem from a systems approach to the urban development process. In short, bearing in mind that the major advantages claimed for the technique have already been enumerated, the primary reasoning behind using an instructional gaming approach can be summarized under the headings of flexibility, transparency, economy and availability. This reasoning is now outlined:

(1) *Flexibility*

The reliance on relatively simple formulations and especially the employment of human agents almost guarantees open-endedness and considerable flexibility. Procedures can be: condensed or extended, stopped and started or even repeated at will, often at modest inconvenience.

(2) *Transparency*

The hybrid approach to simulation through differential abstraction, gross simplification and diverse substitution offers a clarity not readily encountered in the 'real' or 'referent' world. Thus, a means is provided, for example, for bringing out the irrational and for making explicit that which may normally be implicit.

(3) *Economy*
The cost in terms of time, manpower and resources is relatively modest in comparison with other model building enterprises. Real-world mistakes can be expensive as well as dangerous and the complexity of the situation often ensures that what is actually happening is not self-evident, thus a low-cost learning environment for decision-making is preferable.

(4) *Availability*
In coping with dynamic, highly complex and infinitely diverse situations at classroom level, there is often no comparable alternative to gaming-simulation presentation in terms of completeness and bearing in mind the above qualities. Thus, where the ill-defined and seemingly unpredictable exist side by side with clarity and precision in the system under study a hybrid form of simulation may sometimes represent the sole available approach, given such constraints as the present state of knowledge, current resources, and immediate requirements.

Finally, on the question of requirements and the profitable use of games, the discussion must return to the specific purpose of instructional simulation systems and the nature of the referent. Meier and Duke's propositions have served to open the door to a greater level of discrimination here, and it only remains to emphasize that ultimately a more sophisticated rationale must rest upon achieving greater precision in matching the particular problem situation with the appropriate means of approach and 'mix' of resources. Since learning situations differ one from another, this is to suggest that a planning game which is considered particularly adequate in one set of circumstances may well prove to be not as efficient or even totally inadequate in other circumstances. As Meier and Duke (1966, p. 15) have reminded us, the real challenge is in relating needs to human and technical resources and then in exploiting *both* to the fullest advantage.

7. Simulation in practice

In designing and using gaming-simulation models in planning education over the last five years certain significant research and development frontiers have become clearly recognizable; these are now considered first, under three main headings connected with model selection and design, then with respect to operation procedures and finally in relation to the evaluation of instructional planning systems. Then, with these methodological considerations in mind, the chapter is concluded with a review of some of the uses and wider applications of urban development gaming which already appear to warrant greater attention.

MODEL SELECTION AND DESIGN

The starting points when selecting any teaching technique are, of course, what is to be taught, for what purpose, and at what cost. Once these have been clarified, then in theory the potential user of an instructional simulation system should be able to check what is available against his needs and then be in a position to make a selection. In practice, however, matching the would-be user's requirements with existing operational models is a difficult and lengthy process. It requires searching through an assortment of simulation reports, a tedious review of manuals and the perusal of a multiplicity of duplicated notes yet to be assembled into more cohesive operational instructions. In short, at present there appears to be no way of short-circuiting this devious process and such is the quality of much of the documentation that, even after having surveyed the literature, choosing a gaming-simulation is more an act of faith than a judicious selection of a model with the qualities appropriate to specific needs.

Obviously, such a situation does little to advance and promote the use of instructional simulation systems in the study of the urban development process. The writer's 1969 questionnaire survey made it abundantly clear that many respondents were not aware of all, or even many, of the systems available and relevant to planning. The task of selection is far too arduous and even for those with an interest in the subject there is, as yet, little support for their efforts to keep abreast of the rising tide of literature. Thus, as few teachers have the leisure for such time-consuming investigations, there is a clear need for more authoritative and comprehensive information on instructional simulation systems which, at best, could facilitate clear and quick reference.

To partly remedy this deficiency, the writer (Taylor 1969*a*) has proposed a system of classification which endeavours to build on the extensive work already done in the business and management simulation field by such authorities as Greenlaw *et al.* (1962, pp. 270–340), Rohn (1964, pp. 139–77) and more particularly Loveluck (undated). Using this system a catalogue of instructional simulation systems was compiled by the writer in 1969 and is now updated and summarised briefly in the Directory of Simulations presented in Appendix 2. The Directory does not claim to record all games relevant to the urban development process, partly because from the outset attention has been confined to European and North American experience but it does appear to be the first extensive register of its kind and it is already obvious that there is a need for similar more sophisticated reference systems.

In summary, at the moment little is known about the criteria used by teachers to select or reject specific gaming models. It has been shown, however, that assessing what is available by way of urban simulation games is both difficult and time consuming, and it is to be regretted that becoming familiar with what is at hand is so tortuous and perhaps, in many cases, likely to be so unrewarding.

Failure to locate an instructional simulation system appropriate to the user's requirements confronts the teacher with the daunting prospect of modifying or developing a simulation to fit his specific needs. The task is daunting in the sense that, again, little is available by way of practical guidance and, of course, there is no guarantee that any commitments in this direction will achieve complete or even partial success. What is lacking here is a stock of well-documented case studies which illuminate the design process and spell out, step by step, instructions in operationalizing simulation games.

The basic steps in game design in the social sciences have been outlined by many authors in the business studies field including Andlinger (1958); Green and Sisson (1959); Greenlaw *et al.* (1962) and Loveluck (undated) and, in the same area, there are at least two book-long case studies documenting the development of individual models at Harvard (McKinney 1967) and at what is now Carnegie–Mellon University (Cohen *et al.* 1964). Planning education, to date, has been less fortunate in that comparable reports appear to be nowhere near as common or as complete. Meier (1961*a*, p. 242) has set down some general guidance for the social sciences at large but, with the exception of the work of Duke (1964), Taylor and Maddison (1968) and Feldt (forthcoming) there appears to be little of practical significance. Feldt's forthcoming work perhaps offers the most here in that he has devoted three research reports to the detailed description of the methods and techniques for constructing and operating CLUG derivatives simulating specific areas. This eagerly sought after material, when issued, will represent a considerable advance on what is currently available but much work still remains to be done.

So it is that, not only are the would-be users of instructional simulation systems in need of guidance on the steps to take in considering and selecting a gaming model but they also lack a plentiful supply of design case studies to aid the more venturesome in the construction or modification of models to suit their individual needs. Obviously this is an area where some co-ordination of effort would seem to be immediately profitable and some agreement on the definition of a standard methodology for documenting the construction and operational process would probably do much to advance the current state of the art.

Without this support, the problems of construction and modification of gaming procedures in planning education are diverse as well as numerous. This is to be expected as the technique is new and very largely untried in instructional programmes concerned with aspects of the urban development process. To illustrate the nature of some of the difficulties which may be encountered a number of problems are now considered.

First, there are the problems which are common to all modelling efforts. For example, every model is predicated on the assumption that the designer will get more out of his model than he put into it, but the problem is often not what to put in but what to leave out. In Redgrave's (1962, p. 2) terms this is the Achilles heel of any simulation. No substantial body of theory is as yet available to support the gaming-simulation designer by way of guide lines for determining critical parameters. The temptation is to draw upon statistical data at hand, or relatively accessible, to ignore very largely the contentious or that which is not readily quantifiable. Consequently, any model has embedded within its framework the fruits of human judgements. In short, it is a *partial* view of reality which requires the designer's stand, approach and theoretical base to be explicit if the game is to be seen as a *single* representation of the referent. Re-cycling procedures and the employment of new strategies or players can help to provide a wider overall perspective, if alternative representations are not available, but little is known about the 'trade-off' between these devices and the amelioration of the designer's bias. Thus, on the face of it, for some the pragmatic commitments of model builders are just as dangerous as any subjective teaching presentation. However, as pointed out earlier, the values which enter into gaming procedures are at least often more explicitly stated and hence less dangerous than the implicit values innocuously imparted by the persuasive orator.

Second, having said that any model is only as good as the designer's understanding of that which he is simulating, it is important to note that gaming-simulation procedures endeavour to add something by virtue of their open-endedness with respect to the role of the individual, and the level of his contribution within the simulation. Here lies the problem: to what extent should participant behaviour be circumscribed, especially on those occasions when the player is a surrogate for ignorance concerning the unknown, the overlooked or a multiplicity of irrational forces which are difficult to define by alternative means. This leads on to methods for establishing the gradations of simulation appropriate to the nature of the phenomena being studied and the use to which the model is to be put. Meier and Dukes' contribution with respect to the man–machine interface and resultant compromises has already been acknowledged but much remains to be done to sharpen this understanding.

Third, associated with the preceding difficulties there are a whole range of similar ' trade-off ' problems which are central to the model-building process. For example, there are ' trade-off ' problems concernings the relationship between the game and its players; in certain gaming-simulation models the level of knowledge expected of a player may not always correspond to the ability of the student. This dichotomy is well illustrated by the CLUG derivative models of Taylor and Carter which assume a very low level of awareness of the social processes influencing urban development, lower in fact than most planning undergraduates appear to possess. So the use of the game has ramifications for the student who identifies certain shortcomings within the model and has difficulty in accepting these inadequacies. Here, it is important to stress that complete realism and comprehensive coverage cannot be expected. For teaching purposes the model need only possess verisimilitude – the characteristic of appearing real to the students.

Here, a distinction must be made between gaming-simulation *realism*, i.e. the model's correspondence with reality; and with game *verisimilitude*, i.e. the model's ability to convince the participants that the simulation is both real and appropriate. Verisimilitude, however, is often one thing to one individual and often quite another thing to any other. This problem has yet to be investigated in any depth, probably because it represents a research frontier which offers little by way of immediate pay-off. Yet in passing, two remarks seem relevant. On the one hand, verisimilitude is not necessarily the product of a commitment to realistic data – some components *may* have to be overdrawn or distorted in the interests of facilitating student learning and the achievement of narrow teaching objectives. On the other hand, one of the major tasks in game design is the modulation of feedback contingent upon certain actions. In other words, game pay-offs have a pedagogic function in shaping learning, therefore it is reasonable to assume that, in aiding transfer in training, certain contingencies should reflect or parallel those encountered in the referent.

Following on from the problems of realism and verisimilitude, there are the doubts concerning the level of complexity proper to the game. As might be expected there is a natural desire on the part of the game designer to make the game as accurate a simulation as possible. Understandably, students wish to have a game which is as simple as possible; which may tend to compromise the complexity and accuracy of the simulation. The difficulty here is one of achieving a proper balance between extreme complexity and oversimplification, whilst, at the same time, preserving a semblance of reality. No matter how complex the process under study it must be remembered that one important attribute of gaming lies in its ability to decompose involved dynamic operations into a series of very simply expressed actions or sequences of events.

Finally, associated with the preceding difficulties, there are problems of game design which concern quantification. It has been shown that many designers lean heavily towards those elements in the urban environment which can be simply expressed in numerical terms. Social values are sacrificed in favour of the more readily quantifiable and yet, part of the *raison d'être* of gaming lies in its ability to achieve some balance or compromise in the representation process. Debate largely centres on units of measurement appropriate to social exchange and as Hall, amongst others, has pointed out, efforts to resolve this dilemma have not been particularly rewarding. In short (Hall 1967, p. 19): 'Despite considerable philosophical effort since the days of Bentham, the nearest mankind has ever come to a universal felicific calculus is the index of money.' However, there are some indications (Duke 1966 and Gamson 1969) that other measures of value are being evolved in an effort to encompass a wider range of variables in the gaming environment. But here again it must be recalled that, for most phenomena there are many possible game representations and a *single* model cannot be expected to handle all questions asked. The temptation for game designers to cover more ground in greater detail has somehow to be resisted. Unfortunately, the more complex and total a simulation, then the more complex is the equipment needed. Computer technology has already shown itself able to handle some of these problems but often at a high financial cost and with a loss of immediacy. Thus, with respect to the foregoing construction problems, it must be acknowledged that little is known about the overall design process and, in particular, considerable uncertainty surrounds the 'trade-offs' between model accuracy, verisimilitude, cost, 'playability' and tolerable, as well as appropriate, degrees of abstraction.

OPERATIONAL PROCEDURES

The operation and control functions of the gaming process are closely interwoven, consequently any discussion of the on-going mechanics central to instructional simulation systems must bring in certain evaluation procedures. However, this section will endeavour to concentrate on that part of the sequential decision process which relates to student engagement and on the research opportunities presented by this component of the didactic system.

To review what progress has been made by urban research in this direction is not a difficult task as so little appears to have been done. For example, some evidence has already been presented of a few gaming-simulation models which have led to preliminary experimentation with a combination of professional and student manpower. In addition, the writer's 1969 utilization survey revealed: a small number of administrators who have started to consider the effect of time pressures and other modulations of play on participant behaviour; others, who have started to study the development of organizational groupings and communication flows; a few instructors who have systematically organized game role-reversal opportunities; one institution which has sought to link sub-games to a main game and another which has made conscious efforts to combine gaming models with related techniques such as role-playing and in-basket exercises. But, as yet, these efforts are extremely tentative and a total awareness of the potential of complete instructional simulation systems is perhaps only just beginning to emerge.

In creating this appreciation of the total system, the teacher's role is fundamental. It is his task to co-ordinate all system components so that, throughout the experience, the student is able to review systematically what he has or has not learnt and so correlate this with what he was expected to learn. Therefore, it is incumbent upon the teacher to put each element of the instructional simulation into its proper perspective so that as many actions as possible are sensitively inter-related in response to the changing needs of the individual, the group and evolving environment. At a specific operational level, for instance, he must indicate to the players how, and when, certain actions violate specific theories and, conversely, how and when events take place in accord with theory. In sum, the teacher is a safety net or adjustment device between the system and the individual, he is a final check to ensure that erroneous learning is quickly corrected and that the lessons learnt throughout the experience are adequately reinforced,

Thus, to maximize on the learning experience, the teacher needs to know much more than at present about the learning process relevant to instructional simulation systems. From the outset, it would obviously be beneficial to know more about why gaming-simulation participation appears to generate such high levels of motivation and involvement. Partial explanations may be found in: the common desire to out-perform colleagues; the general wish for competition; the challenge to examine and master the unknown; the freedom from instructor–student polarization; and the degree of stimulation given to all of the five senses, but it must be acknowledged that such a partial listing is no more than speculation. Regrettably, it remains to be established why games in general hold an intrinsic interest for so many and why specific types of gaming procedures hold particular interest.

Similarly, the role and development of the behavioural component in games is almost unknown. Little research has been done on: the derivation of learning, the acquisition of skills, and endurance of knowledge relevant to the individual or the group in instructional simulation systems. For example, Taylor and Carter (1969*b*) have shown in Figure 8 that certain levels of interaction between four parties – the models in the game, the teams, the player and the operators – are typical of gaming-simulation models and yet, how to attach values to the learning derived from each of these in specific contexts is, as yet, in no way clear. As the figure shows, not only do players, as a team,

Figure 8. *Levels of interaction in a typical gaming-simulation model*

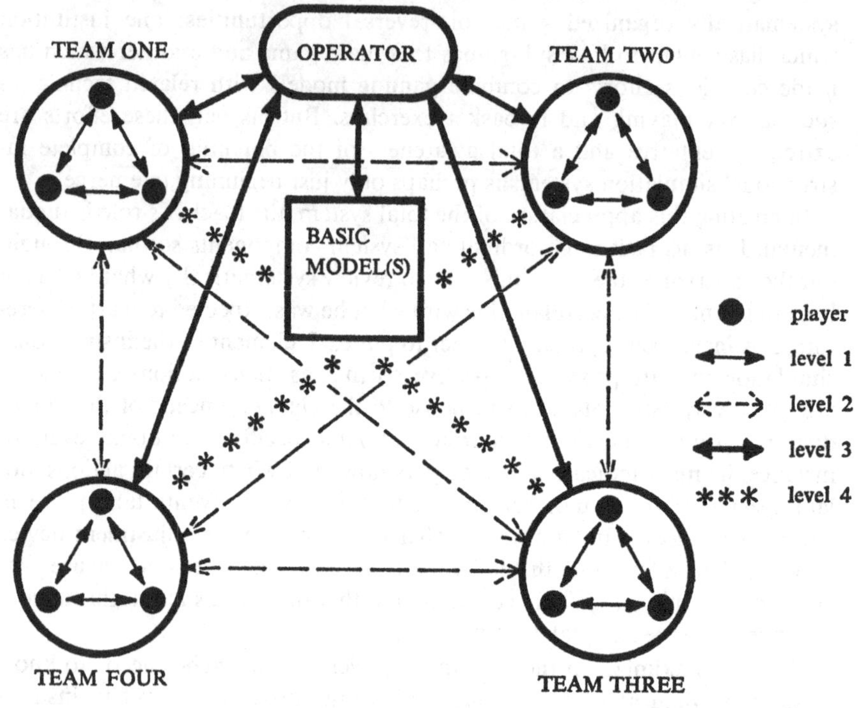

Source: Taylor and Carter (1969b, p. 23).

interact with the model, but other informal learning opportunities are presented as player interacts with player, team with team, and the teams with the operators. In each of these situations, players have potentially something to discover and it is this range of opportunities which has yet to be brought out and fully understood.

As Taylor and Carter (1969*b*) have reminded us, a fundamental educational approach is one of guided self-discovery; one teaches students by laying down situations from which they discover new rules, theories, principles about the real world. There are two basic inputs then into an educational situation, given the objectives: the knowledge students already have and the structure given to the situation by the teacher. One then works on the assumption that student understanding grows incrementally; as a student familiarizes himself with a principle, discovers its dimensions and applicability, so he adds it to his store of knowledge onto which fresh knowledge is grafted, as the situation is explored more fully. The game model has a structure defined by the teacher but the operational simulation is only partly defined and controlled by the teacher. Opportunities for informal learning are presented, as outlined in Figure 8, and it is a greater knowledge of these opportunities which is now required.

In particular, Walford (1968, p. 36) has gone so far as to suggest that gaming-simulations might well provide a useful vehicle for studying the hypothesis that students often learn better from their contemporaries than from their teachers. Certainly, games appear to do much to encourage such learning and this represents but one aspect of player performance which warrants investigation alongside studies into how learning varies from individual to individual, dependent upon: their background; their role; the size, composition and organization of their team; the tempo of the game and the length of participation. Similarly, little is known about how environmental relationships, in terms of the shape, size and disposition of facilities, affect player performance.

The opportunities for those with particular interests in human relations, intellectual development and organizational behaviour are such that it seems unlikely that the game, as a social science laboratory, will continue to be neglected by practitioners in these areas. However, for the urban studies specialist, it appears that most interest will centre upon the model building process and the control mechanisms held by both the designer and the administrators. On the latter point, the 'pay-offs' or feedback functions obviously require much greater study, especially if the teacher is to fully appreciate how these elements shape learning and subsequent behaviour. The part stress, competition, and conflict play in the gaming process particularly deserve to be investigated as do the characteristics relevant to successful participation in prescribed conditions. Each of these areas is an important variable in determining the efficacy of the technique and, until their relative importance is established in set situations, the teacher is, of course, considerably handicapped.

EVALUATING THE EXPERIENCE

At the outset of any discussion on this topic, it is important to recognize that to determine the effectiveness of *any* teaching technique is a difficult task. What may be an effective technique in some situations, in the hands of some instructors, or with some students, may be quite ineffective in other situations or in the hands of other teachers or with different students. From this, it seems likely that, in place of a single overall indicator of efficacy, there are likely to be many factors which have to be considered in what might best be termed a multi-faceted approach to evaluation. Here the resultant short term research strategy would be aimed at assessing what aspects of a particular game or instructional simulation system appeared well suited to which situations as opposed to initiating any attempts to assess overall validity or efficacy.

The problem here, as Raser (1969, p. 138) has pointed out, is the lack of adequate measurement devices. Even if one accepts a multi-faceted concept of validity and examines several types of validity relating, for example, to validity of method or results, the yardsticks with which to tackle this problem are not apparently to hand. To date, only Duke and Monroe appear to have made any progress with regard to urban gaming systems and their evaluation studies, although narrow, may have already gone further than any other published work relating to the use of other teaching techniques in planning education. If further progress is to be made at least two sorts of initiative seem to be required. First, more extensive enquiries are needed to establish reasonable statistical comparison – obviously, firm conclusions cannot be drawn from evaluating one teaching method against another with limited numbers, on one occasion, in a single department. Second, an increased number of different orientations or approaches to assessment are required in sympathy with the multi-faceted concept of efficacy just propounded.

With these thoughts in mind, there are several distinctions which should be mentioned in connection with the evaluation of participant performance and the determination of game ' pay-offs '. As with the preceding discussion and, as in life itself, there are few single yardsticks with which to monitor participant performance in the synthetic environment. Not only are there few yardsticks but, as the military strategists have found out to their cost, when searching for increasing precision in their war gaming endeavours, there are considerable difficulties in expressing wins or losses in numerical terms. This is perhaps how it should be, for the real world offers a range of ' pay-offs ' in such varied terms as power, public prestige, self aggrandisement, happiness or financial success. However, the game world tends to be more prescribed and success is generally established in terms associated with individual assessment, team opinion and instructor appraisal. Certain ' pay-offs ' are distinct products of data and value judgements embedded in the model; others are determined by the interplay of the behavioural components in relation to each other and their environment. In arriving at a balanced assessment of participant performance, all these indicators have to be taken into account. The point being made here is simply that evaluation of performance is not only a question of how quickly a player finds out how to master the parameters of the model as well as the vicissitudes of other decision-makers but other subtleties of performance are of interest. In short, the criterion of

Figure 9. *'Semantic differential' scaling assessment form and a sample set of rating profiles*

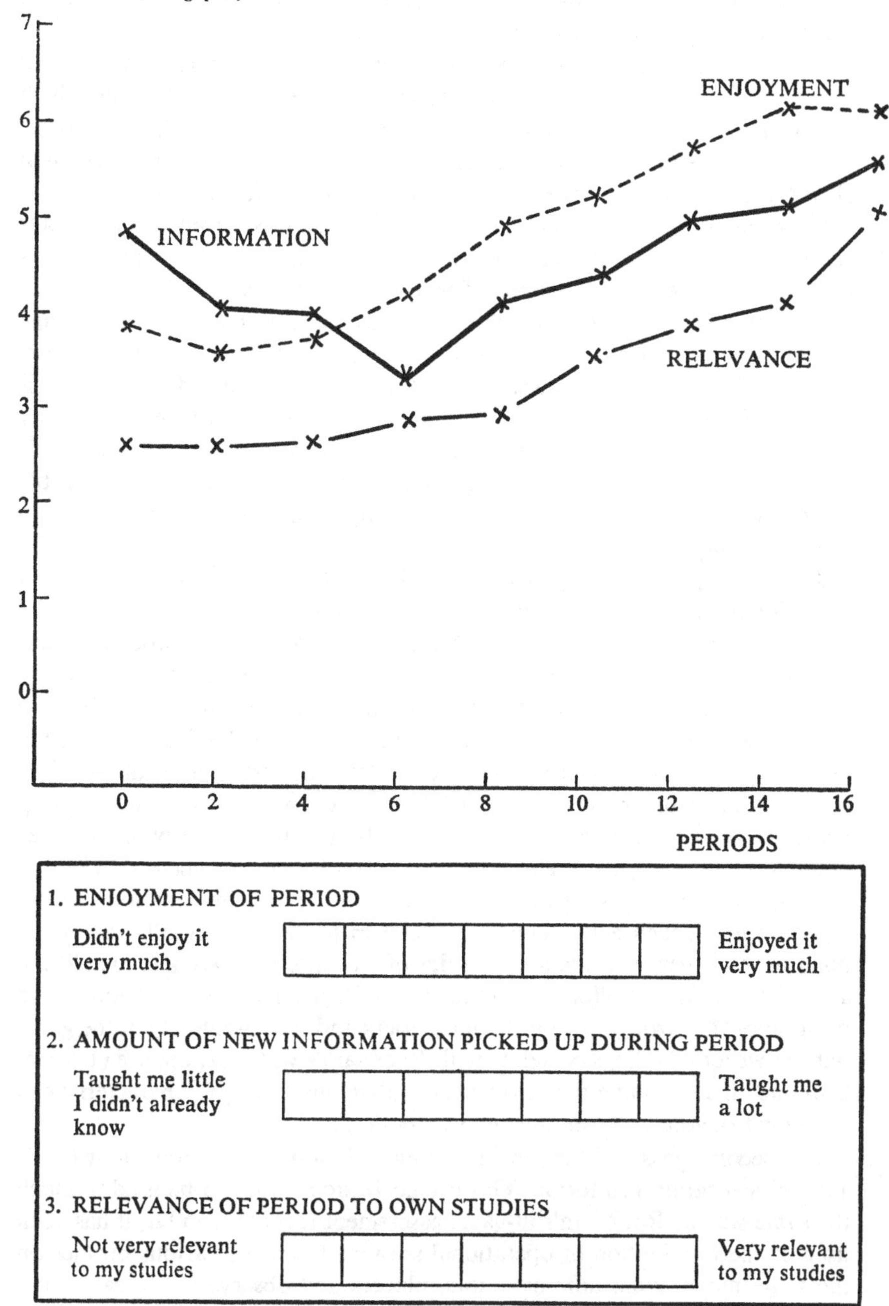

1. ENJOYMENT OF PERIOD

Didn't enjoy it very much								Enjoyed it very much

2. AMOUNT OF NEW INFORMATION PICKED UP DURING PERIOD

Taught me little I didn't already know								Taught me a lot

3. RELEVANCE OF PERIOD TO OWN STUDIES

Not very relevant to my studies								Very relevant to my studies

Source: Compiled from period returns from an interdisciplinary group of 14 undergraduates using the LUGS model under the author's supervision at the University of Sheffield—July 1967.

participant success is in no way simple or straightforward and progress is again likely to result from drawing upon an increasing number of different orientations to the same problem.

By way of example, it is perhaps relevant to quote from the writer's own experience on but two of the very practical assessment techniques which might be immediately considered for wider field testing. One of these methods is a simple monitoring device developed and used by Rackham (1970), a psychologist at the University of Sheffield. This most effective self-scoring record consists of measurements made on three seven-point bi-polar scales which respectively relate to game enjoyment, relevance and information content as set out in Figure 9. The approach derives from Osgood *et al.*'s (1965) development of semantic differential techniques and has the advantage of being simple, unobtrusive, quick and, above all, cheap. The session assessment blank shown in the lower half of Figure 9 is completed by all participants every decision period and they are then collected and scored as play progresses. From these ratings, a total awareness of certain participant attitudes can be continuously built up which have relevance on the desirable length of play if breakpoints are to be pedagogically as well as psychologically satisfying.

The technique can also be employed as an aid in the actual modulation of operational sessions. Here, the rating profiles serve as a guide when the operator wishes to vary the length of decision periods or to expose particular individuals to specific roles or testing situations. The feedback achieved through the assessment form is both clear-cut and immediate and it is but one measure whose value can only be determined over time. To date, perhaps the most significant characteristic of the writer's assessments with Sheffield undergraduates using the LUGS model is the ' trough ' effect relating to group perceived enjoyment and receipt of information as illustrated by the profiles in the top half of Figure 9. Here, a critical point is reached when, after one or two introductory decision cycles, both enjoyment and information perceptions seem to reach a low ebb. This appears to be a potentially disruptive point in the experience and a knowledge of its existence does help to alert the administrator in his efforts to counter or mitigate its effect. Obviously the position of the ' trough ' varies with the users and the complexity of the game but the writer's findings correlate with Rackham's wider experience (1970, p. 206) and there is some reason to believe that this ' trough ' characteristic is common to a good number of gaming activities.

The second possible alternative scaling technique stems from the writer's 1969 classification pro forma. This categorization form can be used in much the same way as Rackham's in-game assessment form but, so far, it has been used at the termination of operational sessions to record various profiles on the same model from administrators, players and observers. In each case, the perception profile is compiled from the identification of certain of the listed characteristics on the form and from the individual's assessment of the extent to which these properties are present in the game. If these subjective ratings are not in accord with the administrator's objectives he can then attempt to plug the appropriate gaps through his handling of critique sessions or through further gaming sessions. Feedback of this kind is often painful to the instructor and an experience, hitherto, often regrettably restricted to the rare occasions when a lecturer examines a student's notebook, without warning at the end of a lecture.

It would be both presumptuous and premature to over-emphasize the importance of such gross measurements and yet it is felt that increasing discernment might well spring from, amongst other things, related forms of such types of scaling. For example, and following from the above discussion, a new and improved classification system could bring with it a better and more precise tool for game assessment. This would take its place alongside other means of evaluation each providing a specific insight into the instructional simulation system's value. A multi-faceted approach to efficacy could well rest on the correlation of a large number of wider ranging assessments and, in such circumstances, the present modest steps are but minor tentative probes in process of refinement. As Rackham (1970, p. 210) has pointed out, they do not pretend to answer global questions about effectiveness but they do offer some direct practical help which appears to have immediate relevance. It is at this level that much work is urgently needed.

The barriers in the path of progress are legion and, if gaming-simulation is to pass from art into science then the divisions between education, psychology and the subject specialists' preserves will have to recede. Perhaps problems here can easily be over-emphasized and the exciting opportunities under-emphasized. Consequently, to indicate the nature of what lies ahead, some factors which complicate the process of evaluation are now briefly described.

At the operational level, one of the major problems in trying to assess the achievements of an innovatory educational method is the 'Hawthorne effect'; that is the extent to which the group which is the test bed for a new idea reacts not only to the new idea but also to the extra energy, concern and surveillance which is brought to bear on them. In short, the term derives from the Hawthorne Research Studies carried out in the Hawthorne Plant of the Western Electric Company in Chicago (Mayo 1957). In this study of productivity, progressive increases in the amount of illumination of the work place led to increases in productivity, changes in the organization of work breaks produced similar results, and then finally with the situation back to the original state, even higher production was achieved. At least two things were revealed, one clear, the other less so. What was clear was that it was extremely difficult to isolate any one environmental element or any single behavioural component in the process of increasing production. New methods produced higher productivity, but further increases could be attributed to the strong group relationships established as a result of the study. This brings up the less clear, but intriguing, point of what influence the research had on the work people in achieving higher production.

So it may be that some of the success of gaming derives from the 'Hawthorne effect'. For, as students like to think that teaching content and methods are constantly under review for their benefit, and the demonstration of this through the introduction of a new and relatively unknown technique excites interest, encourages a high degree of involvement and participation, then perhaps, in turn, this helps to raise the students' level of achievement. In passing, it could then be argued that innovation in teaching content and method in itself is useful if the result is increased achievement. Be this as it may, the point to note here is that any assessment of the achievements of gaming-simulation must take into account the 'Hawthorne effect' and from the measurement point of view, attempt to neutralize this influence. So far, it must be acknowledged that current testing in urban studies has not yet reached this level of sophistication and this type of examination represents yet another of the challenges to be faced by the teacher using instructional simulation systems in the future.

The actual nature of the gaming process also brings with it particular evaluation difficulties as not only do games tend to emphasize the transmission of insights or understandings as opposed to facts and skill but they also involve a rich interplay between participants which means that it is often difficult to disentangle separate individual actions from the performance of the team or entire teaching group. Thus, the more familiar and straightforward educational testing procedures are likely to have to give way to more elaborate and hence more expensive assessment methods. An indication of the new analytical approaches being evolved to meet these difficulties is to be seen in the work of Gray (1970), Paterson (1970), Domitriou (1971), Mitchell (1971) and Wiener (1971), using sound and videotape recordings.

In addition, other more intractable problems remain and one of these certainly relates to the fact that learning transfer is by no means automatic. Consequently longitudinal studies are required to indicate the carry over of knowledge, skills and understanding over time. That is to say that the extent to which the participant subsequently applies what he has derived from the gaming experience has to be determined. This is, of course, one of the Herculean tasks in higher education which has attracted, as yet, little attention.

Lastly, when discussing the particular complexities attached to evaluating instructional simulation systems, it seems necessary to consider the difficulties arising out of the rapidly developing state of the art. The concentration on the design of new models, coupled with the continual search for improvements to existing games, has directed greater attention to opening up future instructional possibilities as opposed to carrying out appraisals on the intermediate achievements of present didactic systems. This emphasis not only produces a labour problem but, in addition, does much to frustrate the efforts of would-be assessors as far as possible wanting a finite object to evaluate over a specified period of time. Instead, there are at least three evolving systems which should be the subject of examination: the basic gaming model, the procedures involved in the instructional simulation system, and the total curriculum. Each one cannot be viewed as an isolated component standing or falling by itself and each one cannot be considered as a static element as all three are open to continual modification as more experience is gained and teaching objectives are refined.

In summary, at least two things emerge from this discussion. One, the various evaluation procedures, already described or known to exist, have yet to be applied widely or rigorously in practice. To date, impressionistic evaluation experiences and the early related appraisals are not sufficient to mitigate against minority or chance elements, nor have they reached a statistical sophistication sufficient to enable them to tackle such problems as the 'Hawthorne effect'. To some extent, it is hoped that these shortcomings can, in due course, be partly ameliorated by repeated plays amongst a wider circle of participants and institutions. Second, in line with the propounded multi-faceted approach to evaluation, new tools of measurement may well need to be fashioned to help meet the particular requirements of instructional simulation procedures.

Finally, to establish a better perspective on the state of the art in its infancy, a comparative comment has to be made relative to evaluation in higher education as a whole, and in relation to other users of gaming-simulations, in particular. In brief, the progress achieved in evaluating urban development games, in a relatively short period, although modest, is in no way overshadowed by the results of attention devoted to gaming techniques in other, more developed fields *or* by the findings of research into other teaching methods at university level. Indeed, there is good reason to believe that in planning education the assessments of gaming-simulation procedures have already advanced further than appraisals made in connection with any other teaching techniques used in this area.

AREAS OF APPLICATION

So far this chapter has outlined certain methodological frontiers in the hope of promoting wider discussion and increased experimentation in connection with the developing technology of instructional simulation systems. Now, with similar objectives in mind and having set down certain methodological frontiers, the purposes and wider applications of the technique may reasonably be considered. Thus, the rest of this chapter is devoted to mapping out some of the more promising uses of gaming-simulation procedures which might profitably be examined in attempting to improve and enlarge upon the utility of the technique in urban and regional studies.

The purposes to which games of the urban development process are already being put are in few ways clear and obviously greater precision in this area is long overdue. The writer's 1969 questionnaire survey helped little by way of precision but did reveal something of the nature of current commitments in terms of the developments in hand and of future projects being formulated or considered. Some of these commitments and proposals are worthy of elaboration after outlining how such remarks might best be structured.

The treatment of gaming-simulation has already to some extent been circumscribed by the investigator's decision to restrict attention to the *instructional* uses of urban development games in *higher education.* At this level at least three major areas of application are already discernible with respect to teaching, training and research. These three areas are not mutually exclusive and already games are being used to serve all three purposes. To date, the teaching function, as one might expect, encompasses the most diverse audience; here games are being employed to study a variety of systems having an intrinsic interest to scholars in such fields as economics, geography, law and sociology. In the training situation, games are largely being used by administrators, engineers and planners to allow the trainee to practise certain specified roles thought to be comparable to those encountered in the referent system *or* to allow those already occupying real-world roles to upgrade their performance. The research function is perhaps the least developed and yet there are indications that an assortment of properties particularly relating to model parameters, personal organization, and game interaction have been, or are in the course of being, examined at a fundamental level. This interest, although modest, promises to be one source of improvement in the development of theory and methodology for both education and planning.

Having set down the preceding three broad categories of usage, then gaming-simulations can further be classified in accordance with certain prescribed functional objectives. If, for exploratory purposes, and at a high order of generality, it is assumed that the central purpose of any instructional simulation system is to produce certain desired effects more efficiently than previously then three types of communication objectives are discernible in a form which very largely transcends the three areas of application previously discussed. Consequently, game uses are now considered with respect to efforts to transmit or elicit: *certain information or knowledge*; *prescribed skills*; and *particular varieties of understanding.*

First, taking those gaming procedures which predominantly attempt to transmit information or knowledge, and thus are largely concerned with facts and principles, then the degree of overlap with the other two specified communication areas becomes quickly discernible. In this sense, such models often offer a number of partial answers to certain problems and an appreciation of this inadequacy is brought into perspective by the other components of the instructional simulation system. In short, the game itself is not a general problem solver or a total instructional programme. Furthermore, games in this area can equally well create a ' *teaching* ' as well as a ' *learning* ' environment. For instance, when professionals and students with differing objectives, participate together, the student often receives certain *tuition* on how decisions are, or can be, made by those in authority and the professionals not only bring added richness to the simulation, through their extensive experience, but have the opportunity to *learn* about student generated ideas and knowledge. This facility is more common to games directing their attention to the provision of certain understandings and is less so when the communication of particular skills is at stake.

Simultaneous learning on various levels can perhaps best be seen in early gaming efforts to link the planner and the planned. Efforts in this direction were initiated to acquire information from the consumer as well as to pass on certain knowledge about the planning process so that a meaningful dialogue could be established. The appearance of ' advocacy ' planning (Davidoff 1965) and the movement towards greater public participation in the planning process (Skeffington 1969) are perhaps testimonies to the serious breakdown in communications between the two groups. In brief, the problem can be stated thus: on the one hand, the planner has often in the past tended to look only for public acquiescence for his proposals rather than a measure of real understanding or active support. Whilst laymen, on the other hand, have appeared to lack interest in planned development or have been continually frustrated at their inability to comprehend or follow the ramifications attached to planning projects.

Whatever the relative merits of each case, the two sides were first brought together in the gaming environment by a number of American practitioners (Berkeley 1968, Bourgeois 1969, and Mitchell 1968), apparently motivated by a desire to overcome past shortcomings and establish a closer understanding in the interests of achieving a better creative relationship. More specifically, early games such as Mitchell's URBAN PLANNING SIMULATION and Berger and Walford's TRADE-OFF were conceived as data collection devices for canvassing opinion; their great advantage being that through this form of simulation procedure the participants could unself-consciously express through their actions, as well as words, their attitudes, preferences and predilections about which they might find it difficult to verbalize in other circumstances. In short, the game is one means of narrowing the gap between the planner and the planned and provides a flexible means of contact for frank and immediate two-way expression.

By exploring this relationship further, it is possible to conceive of games which go beyond simple data collection devices and attempt to introduce basic principles and ideas to the public whilst the planners monitor their reactions under varying conditions. Already Domitriou (personal communication) at the Leeds School of Planning has begun examining the reactions of people in the simulation process as a guide to probable reactions in the referent system. This work could be developed in many ways in relation to the performance of individuals and groups, and as a way of bringing out certain characteristics of interpersonal relations. Little is known about how differing segments of the community establish their environmental objectives as well as how they view the process of change. In this context, gaming-simulation could become, as in the Goodman POLICY NEGOTIATION GAME, a means of exercising, on a trial basis, certain information, organization, feedback and control functions of the planning system not only to arouse interest but to demonstrate the virtue and advantage of planning itself. Planners have been slow to attempt to study these topics but, given the present public pressure and the availability of such a tool, it is reasonable to expect a greater commitment in this area in the future.

What has so far been suggested in no way implies that games are the sole means of eliciting information or of communicating with the public. At best they represent one means of *augmenting* information gained from conventional sources and at worst they *may* proceed no further than current games used to introduce the student to his public and employers. The Dundee (Henderson 1968) and Toronto (Benjamin 1968) initiatives in this direction are perhaps the tip of the iceberg and the efforts of the Drexel Institute in Philadelphia (Pennington 1969) may do much to reveal how the potential professional learns about his subject-matter, not to mention how much he discovers about the pedagogic process in terms of learning how to learn. This last point, of course, is vital to any potential planner in times of rapid change.

Turning now to the second area of communication, the transmission and development of particular skills. In this sense, the game environment creates a testing ground, half-way between theory and practice, for the manipulation of certain properties, over time and with the aid of certain tools relative to the decision-making process. Here, at a basic level, the student can be introduced to: the elements in the decision process; the implications of various alternative decision chains; and the importance of certain critical decisions; the objective of such an introduction being to increase the student's analytical skills in particular spheres of activity so that he learns how to: identify sources of information, use certain techniques for processing data derived from such sources and, above all, appreciate the minimum amount of information *relevant* to particular situations.

Opportunities to practise the techniques for processing information in the dynamic setting of a game are again most advanced in the United States. Here, considerable investment has gone into computer technology with respect to developing the man–machine interface in such areas as computer graphics and the construction of automated data banks. Dramatic illustrations of these uses are to be found in the METRO project (Duke *et al.* 1966) and the G.S.P.I.A. MANAGEMENT PLANNING GAME (Hendricks *et al.* 1966). However, it is already obvious that less ambitious models are being developed to effect a relatively painless introduction to electronic computing and data processing at fairly elementary levels. Of vital importance in connection with such introductions is the fact that the student can not only familiarize himself with such techniques alongside his colleagues, but can also *practise* these skills in an operational setting.

The relationship with other colleagues can further be developed at least with regard to the acquisition of two other sorts of skill. First, there are the opportunities presented in the area of social sensitivity training. This is an area not prominent in planning education and yet, in terms of dynamic urban management, the success of group decision-making is to some extent founded upon an appreciation of the working of organizations and the exercise and manipulation of leadership. Second, and closely related to the involvement with large organizations and group activities, are the opportunities presented to analyse and synthesize widely divergent data from multi-disciplinary sources; the objective here being to allow the student to keep pace with the emergence of new and improved management techniques which attempt to draw together problem-orientated expertise. Here it seems that 'Discounted Cash Flow' techniques and 'Planning Programming Budgeting Systems' are representative of some of the techniques with which the planning student might have to become increasingly familiar in assessing the implications of various inter-relationships and alternatives or in channelling numerous efforts towards common objectives.

Finally, with respect to the acquisition of specific skills, a number of games have been designed and are being used to train planning personnel in the operation of certain models. Hendricks (1964) and his associates for instance have constructed a manual game to train the City of San Francisco local authority staff in understanding and manipulating an Arthur D. Little computerized model of the city's housing market. Similarly, Feldt (1967*b*, p. 13) has used his CLUG model as a simple analogue to Consad Research Corporation's elaborate Pittsburgh Community Renewal Model so that a wider audience can come to appreciate how the referent model is operated and how to interpret its output. In quite a different way, Cohn (1968) has produced an ARCHITECTURAL CONTROL SIMULATE to allow potential planners to become more skilled in the management of aesthetic control and in techniques relevant to the planner's interaction with the community power structure. In sum, these are but a few examples of gaming-simulations orientated towards training the planner to become more skilled in improving his performance through an acceptance of wider professional responsibilities and a command of new social technologies.

Third, and last in terms of the communication uses to which games can be put, are those applications which relate primarily to the transmission of certain levels of understanding concerning the milieu within which the planner works. At a general level, for instance, the concern here could be with the 'atmosphere' within which the planner operates, the 'climate' surrounding his activities and the 'empathy' and 'feel' the student seeks to establish before assuming professional responsibilities. Thus, with these sorts of objectives in mind, games seeking to establish different varieties of understanding are generally believed to offer more about recognizing problems than about solving them.

More specifically, there are several examples of how games can be used to bring a number of 'worlds' together through 'tuning' participants into common wavelengths of exchange. At an overview level, potential planners seeking to acquire a synoptic view of the planning scene whilst developing individual interests are continually searching for such simple clarifying mechanisms. P.T.R.C.'s LAND USE TRANSPORT SIMULATION (Macunovich 1967) attempts to present such an overall perspective whilst also attempting to create a greater comprehensive awareness of the multitude of factors affecting official planning machinery. In this sense, the model acts as a frame of reference for perceiving a totality freed from the distractions of background noise.

Similarly, the 'worlds' of practising planners, associated professionals, and governmental decision-makers can be brought together to facilitate greater professional understanding and collaboration. There is ample evidence to indicate poor communications, isolationism and misunderstanding here and the simulation game seems to offer an ideal arena for greater communication and fruitful exchange. For example, on the one hand, professionals concerned with the building sciences often appear to lack an appreciation of governmental decision-making processes and the pressures attendant upon questions of choice, whilst on the other hand decision takers often do not appreciate the professional's position and responsibilities. Gaming-simulation might thus provide the common ground on which members of these bodies might be brought together on much the same lines as major policy-makers interact with tacticians in the political exercises apparently much beloved by the Pentagon. Be that as it may, it does appear that, through joint participation in such games, each group would develop a better knowledge of the other's function and duties. As a result, barriers in communication might slowly recede, individuals and groups would come to recognize what *each* can contribute and achieve as individuals or in concert and ultimately the effectiveness of all disciplines might be enhanced.

One example of this type of very important use is to be found at the Institute of Local Government Studies, University of Birmingham, where Armstrong has designed the ILAG (InLoGov LOCAL AUTHORITY GAME) and the ALEA models partly as a means of bringing together people of differing kinds and levels of professional expertise, both in their own right and in simulated roles. In the games, inter-professional collaboration and exchange of views is rendered necessary, and management skills based on finance are taught in relation to urban expansion problems. Armstrong's simulations, characteristic of many business games, are a test of player ability to harness the endeavours of fellow team members towards a communally defined set of objectives. Thus, one major function of planning games might well be for recognizing the potential contribution of a wider variety of disciplines, and as a basis for organizing disparate approaches to the determination of community objectives, policies, and implementation procedures.

Such interdisciplinary involvement could also do much to make explicit some of the personal and vocational factors in the decision-making process. For example, Greenlaw *et al.* (1962, p. 44) have drawn attention to what has been termed the 'suboptimization problem'. This difficulty exists when objectives are in conflict and when the optimization of one objective results in a lower degree of attainment in all the others. In a dynamic and sequential decision-making exercise such as an operational game, disciplinary or organizational objectives can be inter-related and comparatively evaluated. Thus, acceptable compromises, involving the maximum good and the minimum hardship, can be forged in an environment which has direct meaning to the participant. It is doubtful whether any amount of preaching, in either lecture or books, on the benefits of interdisciplinary team working and on the applications of game theory can have this impact.

Also, in the future, gaming procedures could provide a basis from which the gaps between certain other groups or areas could be appreciated. It seems that differences between national and ideological approaches to planning could be particularly quickly revealed in such a setting and the game would seem to promise much in achieving better vertical integration in the learning process between school, university and research establishments. Last, but by no means least, a better understanding of the relationship between the past, the present and possible futures could result from 'mind stretching' simulations. Here, the problem for the student, academic or practitioner is the development of a transitory tool which frees him from the weight of conventional wisdom and allows him to use the game as an exploratory device to increase his future awareness. In this sense, the game becomes a stepping stone to more rigorous and improved theoretical formulation after certain ambiguities, contradictions and uncertainties have been identified and experimentally manipulated. Players are thus presented with opportunities to live with problems that might not be readily imagined and might otherwise be neglected. In furthering understanding and innovation at this level, perhaps the greatest progress has yet to be achieved, not only with gaming procedures but with all aspects of education.

In summary, then, this chapter has attempted to outline certain research and development frontiers with regard both to the methodology and application of the technique. It seems important to note that this review has been too short to be comprehensive and that the intention has been to present the nature and range of opportunities seemingly offered rather than to identify clear-cut research strategies. Considerable overlap exists from one activity to another and few games as yet appear to be confined to the accomplishment of narrow objectives. It should be evident that the limitations in the current state of the art, as well as the restrictions of space, have forced many of the preceding remarks into the realm of generalization. This predicament serves to emphasize the need for greater discrimination and increasing precision in all future commitments. It is to be hoped that this work will provoke or stimulate such attention and that greater understanding and awareness will spring partly from the writer's efforts to establish what has been achieved thus far and, more especially, what is available to illuminate further practice.

8. The way ahead—emerging trends and future prospects

In concluding this examination of an instructional simulation approach to the urban development process, it seems appropriate to precede any summary observations with two sets of remarks. First, some comments are made on possible trends and characteristics likely to influence the general prospects for work in this field. Second, some notes are presented on a few of the barriers in the path of future progress and on the respective steps which may be necessary to continue and promote further advances. Finally, in drawing together what has gone before, some thoughts are expressed about what might have been revealed and what remains to be accomplished.

THE GENERAL OUTLOOK

In taking a considered view of the ' state of the art ' it seems particularly relevant to be constantly reminded that, with the exception of war games, gaming-simulation is a post-1956 phenomenon and, more importantly, planning games are very much a post-1960 development. Despite the fact that, at present, urban development games are little more than experimental prototypes, the technique appears to be increasingly at the centre of a modest international ' growth industry '. Not only are more games being designed and used year by year but there are ample indications to show that these models are slowly encompassing a wider range of activities and, in many cases, becoming increasingly more sophisticated.

Thus, as Figures 5 and 6 show, one of the prevailing trends in the relatively rapid build-up of interest in instructional planning systems has been the continuous and even growth in the availability of new models. Added to this, if the selection of simulations listed in the directory in Appendix 2 are any guide, the range of game developers, model types and users, is being progressively extended. Such growth has taken place in a decade when considerable attention and funds have been devoted to the application of the technique to military, business and management problems. In particular, much credit must go to the continued ingenuity of the war gamers and the foresight of the American Management Association (A.M.A.) which, in 1957, had the ability and the resources to recognize the potential of the technique outside the military sphere. What might have happened had another professional body other than the A.M.A. been able to visualize these possibilities and been able to employ a similarly talented group of systems specialists to study the application of these same techniques to their particular fields, is an interesting point. Certainly it is worth considering, as only comparatively recently has urbanization and the environment become an increasingly popular frontier and urban development gaming has started to receive comparable treatment and attention. Particularly significant in this upsurge of interest is the very recent transference of the American 'aero-space' expertise into the realm of urban studies and, if results are proportionate to the resources invested, then prospects for continued growth and advancement are extremely promising.

At a far less conclusive level, there are signs of a greater build-up of interest in the production of commercialized gaming materials and 'spin-off' games incidental to other research efforts. On this last point, it is interesting to note that, although the present study is concerned with pedagogic rather than with practical applications of the technique, it is clear that the planning profession is slowly making tentative efforts to develop gaming procedures more appropriate to its vocational needs. For example: Rockwell *et al.* (1968, p. 31) at the Northeastern Illinois Planning Commission have employed a gaming-simulation process to generate alternative sets of decisions affecting urban growth as part of a simulation and evaluation of settlement growth alternatives in the Chicago area; Traffic Research Corporation (1969) have documented preliminary stages in their consideration of gaming systems in connection with the Merseyside Area Land Use and Transportation Survey; and Armstrong and Hobson are currently concerned with developing the NEXUS model into a PROGRAMME PLANNING SIMULATION largely for the Corporation of Liverpool's needs; and Lassiere (1969) and Hoinville (1969 and 1970) have reported on Social and Community Planning Research's continuing efforts, sponsored by the Ministry of Transport to develop an environmental management game to enable people to represent their residential preferences.

Obviously there is likely to be an element of the 'fashionable' in the current enthusiasm for gaming-simulation. The apparent strength of the technique as exhibited by its ability to attract considerable attention, activate quite a number of research efforts and produce such a variety of gaming models can do much to conceal the 'bandwagon' effect on the less scrupulous. Here, the danger lies in games becoming rather superficial status symbols paraded as evidence of the acceptance of curriculum innovation and general progressiveness. This phenomenon is common to every field of endeavour and particularly true of games as, to date, they have lent themselves to 'stunt' presentations and, in the writer's opinion, have suffered because of this susceptibility. Here the real enemy of progress is ignorance and, as always in such cases, the most potent source of improvement is a wider circulation of information.

Setting such problems aside for a moment, there are a number of encouraging signs which warrant description. For instance, in a capsule history of simulation games for classroom use, Boocock and Schild (1968, p. 15) have recognized three distinct (if overlapping) phases of development. Phase I, they have labelled the acceptance on faith period; Phase II, post-honeymoon; and Phase III, realistic optimism. In the urban development game movement it is possible to discern three slightly different stages of evolution. Stage I might be termed the post-1960 slow build-up of interest in the technique; Stage II might be seen as the current trend towards achieving a wider recognition of the potential of the technique; and Stage III might hopefully cover the eventual fruition and multiplication of current plans aimed at more rigorous assessment of theory, experience and validation relative to particular games as used in prescribed settings.

The confidence underpinning such a conception is largely based on: the experience accruing from developments in other fields; the general growth of educational technology; and the associated emergence of commercial learning systems. Thus to elaborate on some of the issues involved in the postulated Stage II, the current trend towards achieving a wider recognition of the potential of the technique, certain planning game landmarks should be identified. For example, Feldt's efforts to make kits of materials readily and widely available deserves special mention; for, as early as 1966, Feldt (personal communication) prepared over twenty trial kits of CLUG primarily to determine whether it was feasible to produce such materials commercially. Encouraged by an exceptionally quick disposal of this stock to both universities and planning offices (kits sold at a subsidized price of 50 dollars a set) a commercial interest successfully took over the mass production of these game materials. Subsequently this group, System Gaming Associates, offered simulation computer programming assistance as well as consultant demonstration services for CLUG, its elaborations, and potential applications. Meanwhile, Michigan State's Urban and Regional Research Institute, under Duke's direction, produced a 'packaged' introduction to CLUG and, in due course, the Washington Center for Metropolitan Studies established a gaming laboratory to provide, amongst other things, a counselling service to integrate simulations into educational programmes and to relate particular models to specific locations. This trend towards organizing and systematizing learning is by no means particular to planning education, in fact it represents a much wider movement, but the significant point is that the movement had its entrée into planning via the development of instructional simulation systems. It would be premature to read too much into isolated events if they did not correlate with activities on a much broader front; however, it is felt that these avenues for extending the use of gaming procedures are indicative of some of the exciting opportunities offered by a fuller exploitation of educational resources. In short, Stage I in the evolution of urban development gaming is being propagated in part by radical means, and the writer places considerable faith in the potential fruitfulness of such approaches.

To counteract any belief that these movements will result in a total mechanization of education and a consequent depersonalization of learning, it is significant to note one clear trend in the rise of urban gaming activity. This trend relates to the level of questioning and personal commitment which has surrounded and become part of many new gaming models. Few institutions have accepted particular games in their original or 'market' state but have used the model as a point of departure and as a basis for future work to suit their individual requirements. Once the theoretical concepts and administrative procedures of the 'raw' model have been commanded then they have often seen the way open to develop modifications and evolve their own derivatives. One of the most beneficial 'pay-offs' in this area, stemming from the unit construction of gaming procedures, is beginning to be seen in the way both staff and students have started to simplify or build on instructional simulation systems for wider exploration or for more varied experimentation.

This interest has in turn given rise to a proliferation of developments which have previously been referred to as 'in-house' or 'institutional' games. In channelling their efforts in this way, planner educators are following their counterparts in business and management studies where the trend has been away from abstract models of generalized situations to a greater interest in games designed to deal with specific problems of particular locations (Kibbee *et al.* 1961, p. 168). The trend in planning is particularly well identified by the extensive adaption of CLUG to various regional or national settings and by the growth of commercial agencies to promote this sort of transference. Of considerable interest to this discussion is the fact that the raw materials of the educational innovators and technologists have not been accepted at face value as tools for immediate universal use; instead, it appears that they have provided, almost without exception, a stimulus for fresh initiatives to fashion instructional simulations more appropriate and directly relevant to particular local needs. This readiness of teachers, not only to experiment with new techniques but to inject their own new thinking into the evolution of instructional simulation systems, obviously augurs well for urban development games if they are to maximize their full potential in numerous settings.

In sum, the burden of this study has been concerned with an approach to urban problems which is slowly crystallizing – by any standards, gaming-simulations are in their infancy and comparatively speaking, urban specialists have not been, as yet, in the forefront of gaming research or development. However, despite the fact that urban development gaming experience is modest and restricted very much to the latter years of the past decade, the number and nature of the models produced during this time, plus the enthusiasm and vigour associated with work in hand, make it difficult not to believe that the quality and coverage of this activity will continue to improve and increase. Certainly, an interpretation of past history and current trends do not point to any other obvious conclusions.

PROBLEMS, OPPORTUNITIES, AND THE FUTURE

This promising picture should not lead the reader to suppose that, at the moment, urban development gaming lacks contention, difficulties, or room for improvement; such a novel approach generates problems of its own and as this examination has attempted to show, there are numerous gaps in the planner's knowledge of the technique. In short, the technique is neither a paragon of virtue nor an exclusive path to a better understanding of urban phenomena. However, it is believed that instructional simulation systems founded on gaming models have some merit and have supplementary relevance in urban studies. With this confidence, the problems that lie ahead can be regarded, not as stumbling blocks, but as springboards to greater progress. To illustrate the nature of some of these difficulties, and of the opportunities offered, a number of these issues are now considered.

Varied experience with the technique has demonstrated to the writer the necessity for two sorts of broad initiatives which might be undertaken to overcome some of the current problems. These initiatives are both overlapping and mutually interdependent, and yet for clarity they will be discussed separately. The first relates to certain steps which could considerably improve continuing efforts to strengthen *linkages* and *exchanges* between all instructional simulation users, or potential users, irrespective of discipline or academic level. And the second type of initiative is concerned more with some of the measures needed to promote the unified development of particular instructional simulation systems and the more precise identification of the most *effective* and most *efficient* solutions to pedagogic problems.

Before reviewing each of these initiatives in turn, certain constraints restricting progress should be briefly restated to provide some perspective to the ensuing discussion. With the present educational planning resources, it is plainly impossible and indeed perhaps extravagant, for many institutions to commit themselves to an extensive programme of experimentation, in this field. The technique is still at a rudimentary and untried stage, and with this in mind the capital costs, in terms of 'programmed kits' and 'packaged' models, are in themselves probably beyond many individual departments. In addition, there are opportunity cost problems arising out of a shortage of staff and time as well as the closely related training difficulties, noted by many respondents to the writer's 1969 survey, arising out of current staff aptitudes and inclinations. On another plane, there are difficulties resulting from a lack of a systematized and rationalized communication network between all those engaged in gaming-simulation research. Here, today's informal unco-ordinated efforts probably allow much effort to be duplicated and restrict lessons being learnt from others in the field. Few games reach the outside world via the publishing house, and the duplicated notes, sometimes freely circulated, are rarely in a form which many would widely welcome.

In sum, conscious efforts have to be made to transfer the expertise at the finger tips of a very few innovators to the finger tips of the body of the educational profession. Simultaneous with this movement should be a parallel effort to ensure that more educationalists are involved in fashioning and evaluating improved means of learning. Thus, in this context, this means focusing greater attention on the development of gaming-simulation procedures in a common effort to make them, not only more readily accessible, but also cheaper, more flexible and generally more efficient. The gulf between pioneer and practitioner has to be further narrowed if both increasing recognition is to be accorded the technique and a growing understanding forged outside a very limited educational circle. It is believed that this publication has shown that, at least, a bridging point *may* have been established and the steps which follow are suggested as a means of strengthening this bridge.

When considering strategies for propagating gaming ideas and new simulation thinking it seems obvious that some form of overall co-ordination is required. The ultimate in this direction would be the establishment of a National or even International Centre for Simulation Studies. This is not beyond the bounds of reason, even for the United Kingdom which already has a National Council for Educational Technology and at least two university centres with special national responsibility for promoting programmed learning and computer technology. One of the primary functions of such a centre, with respect to instructional simulation systems, would be to handle and improve arrangements for collecting, summarizing and circulating gaming-simulation expertise. What is urgently needed is a *recognized* clearing house or information centre whose duties would include: co-ordination of present *ad hoc* efforts; organization of research and the dissemination of results; facilitating a closer working relationship between researchers and the interested practitioner; and above all, the provision of data upon who is doing what, in which situation and to what end.

Complete up-to-date records of existing materials are essential to continued progress in this, as in every, field. The prototype classification system and cataloguing done by the writer have been seen as one move in this direction. Further advances will to some extent rest on a wider comparative assessment of recording methods and information systems; with respect to the latter, much work on an agreed glossary of terminology is required before the feasibility of alternatives becomes clear. The emphasis here ought properly to be directed at putting into every teacher's hands the means for choosing between an abundance of media and technical resources as befits their needs. The enormity of this task means, of course, that it cannot be left solely in the hands of a national body but will require full participation and partnership between *all* those concerned with improving the effectiveness of learning.

To some extent the way ahead, in this respect, has already been highlighted by the emergence, particularly in the United States, of a number of commercial or non-profit agencies offering this type of service. The spearhead of their approach has been the production and dissemination of multi-media kits or study units which aim to supply, as far as possible, total simulation learning systems. These comprehensive ' packages ' have been a very real basis for maintaining and increasing the momentum of work in progress in so far as they have delivered to numerous consumers a palatable means of familiarizing themselves with the expertise of the few in what might best be described as a ' do-it-yourself ' approach to educational ' retooling '. It should be appreciated that gaming systems are particularly amenable to such a form of presentation not only because of their unit construction but also because many instructional simulation systems are in fact often compendia of games, each of which can be used separately or can be plugged into the others to make a coherent whole. Thus, the total model becomes the ultimate in sophistication and, by building in one sub-model at a time, a graded progression of complexity can be slowly achieved. It will consequently be surprising if more of these ' packages ' are not forthcoming and if a greater emphasis is not placed on the production of total simulation learning systems.

Some of the commercial and non-profit agencies have made highly commendable attempts to integrate their study units into extensive curriculum-development projects and, in safeguarding this investment against the pace of increasingly rapid change, they have promoted new training courses, offered considerable demonstrational assistance and provided a full back-up service to up-date materials as well as to supply and replenish equipment stocks. This is not to suggest that the interests of educational commerce always coincide with those of the consumers. The value of some instructional simulation systems has been vastly oversold and the understandable eagerness to capitalize upon the exploitation of novelty has not been without its drawbacks. However, on balance, such initiatives are to be welcomed and it is to be regretted that, with the possible exception of the American Association of Geographers (1967), these revolutionary avenues for fostering innovation and cultivating new knowledge have been neglected by too many urban educationalists. Certainly these methods are likely to be too important in the future to be left largely in the hands of educational commerce.

Finally, with respect to the steps necessary to propagate a wider educational dialogue with regard to simulation systems, it is gratifying to record one concrete development which is already under way. For in 1970 The Sage Corporation commenced publication of a new quarterly journal devoted exclusively to social science simulations. This decision to some extent represents the coming of age of the technique in this area and such a specialized journal should do much to provide a forum for all those working with, or wishing to work with, social simulation systems. In sum, such an instrument should hopefully become the ears and mouthpiece of international exchange helping to ensure that a wider, continuing, and more informed discussion takes place.

Now to turn to some of the measures needed to promote the development of particular instructional simulation systems and the more precise identification of the most *effective* and most *efficient* solutions to pedagogic problems. In the first place, a new attitude to the continuous evaluation of teaching methods must be created. Evaluation and feedback are basic to the ' systems ' approach to the learning (and planning) process; and to be clear about what has been achieved, must, by the nature of things, require an increasing precision in the formulation of needs and objectives. Few would deny that planning education could benefit here.

At the same time, priorities have to be established and carefully planned sequences of research into urgent problems have to be undertaken by concerted efforts to ensure the general statistical validity of findings. The Hale Committee's (1964) call for national and inter-institutional collaboration and exchange is vital here if educators are to start to appreciate the situations and circumstances in which certain techniques are most effective. Such co-operation is also vital if a better cost-effectiveness ratio is to be achieved. For only if the technique is used with a large audience on a co-operative basis will pay-offs, in terms of staff and equipment economies, be fully realized. In this respect, it is by no means accidental that certain centres of excellence have emerged and particular groupings of institutions have occurred. In both cases, inter-departmental relationships have been stimulated by the limited resources available and the common desire to build up greater overall expertise at a quicker rate. Certainly it seems that more joint projects and greater inter-use of staff and equipment will be essential if institutions outside the United States are to approach the American levels of investment.

The costs, the shortage of models, and the limited availability of expertise all demonstrate to the writer the clear necessity for increased usage of what is already on hand if economies are to be achieved immediately. The number of games available, the number of institutions using them, and the amount of time and money invested in design, development and administration at first sight might be considered to be relatively small, and yet, in planning education this commitment must be regarded as substantial. Certainly, few other educational techniques used in this area have received so much recent attention, if one is to judge by an analysis of the literature. This fact reinforces the writer's concern about the encouragement of wider co-operation if better results are to be more quickly forthcoming and if the profession is to cut down on the wastage of time, money and human resources which too commonly exists, at the moment, in connection with the majority of gaming endeavours.

The problem of cost effectiveness is by no means an easy one to overcome. Obviously the economies suggested thus far are more in the order of a simple rationalization of existing practices rather than fundamental steps designed to ensure more penetrating cost consciousness. If such steps are to be taken, then there are at least two stumbling-blocks which will have to be surmounted. The first relates to the difficulty of determining a comparative cost criteria system by which one method can be judged against another in answering the question: what resources are required and what advantages are offered by, for example, technique *A* as compared with technique *B* when attempting to satisfy the same educational objectives. The absence of a more highly developed science of educational measurement does not lead one to conclude that much progress will be achieved on this front in the near future.

The second stumbling block to overcome follows on from the previous point and relates to the problem of identifying the role of the individual and the steps necessary to isolate his significance in comparative assessments. For example, with reference to the personnel involved in gaming research and development, it has already been pointed out that most reports on this subject originate from authors who have invested heavily in the technique. It is equally well known that almost any teaching innovation works well if taught by its creator or by the young and energetic, idealistically searching for greater success in striving to keep up with the pioneers. What is not so clear is how innovatory techniques work out in other hands *or* conversely what would be the result if the same enthusiasm and energy were applied to traditional methods. In short, to what extent does the effective use of any method, old or new, depend upon the quality of both the instructor and the students. This question may never be adequately answered but a balanced proportion of effort devoted to comparative field-testing of games outside their originating institutions, away from the heavy commitment of the designers and in differing teaching contexts, might do much to enlarge the instructor's understanding of this notorious gap in the educationalist's knowledge.

As with any new techniques, misunderstanding and abuse are always possible. Excessive claims are just as harmful as unjustifiable criticism. As in all things, discrimination will have to be acquired and this can only come from systematic appraisal and objective evaluation. Any discussion of new techniques should not become a question of the polarities for or against but rather a question of degrees of fitness, relative to prescribed purposes; just as the debate surrounding programmed learning no longer seems to centre about whether such learning is intrinsically good or bad but rather whether a *particular programme* has a proven efficacy in *set situations*. This, of course, is easier said than done, as many psychologists attracted to programmed learning have testified (see, for example, Borger and Seaborne 1966). Furthermore, in the planners' search for means of distinguishing between good or bad instructional models to fit their particular needs they have not, as yet, had the same degree of support from educational psychologists. Thus the real issue rests not with the overall validity of simulation as a learning technique but upon the *proven* educational relevance of particular simulation procedures in specific settings.

At this discriminating level, the case for using certain games in prescribed circumstances stands very largely unproven, as does the case for so many teaching techniques used in particular settings in higher education. Indeed, with this in mind, it may be considered that the lack of accurate gaming-simulation assessments stems partly from the general poverty of information on the efficiency of commoner teaching techniques. The older techniques have withstood the ' test ' of time and perhaps little else – newer approaches have to do more. Yet it is believed, as already stated, that gaming validation studies, however modest, have already gone further than any other methodological appraisals in planning education. This progress is by no means sufficient to justify complacency as much work obviously remains to be done, but it should not be forgotten that a start has been made and initial results are encouraging.

Few, if any, games claim to be sufficient in themselves and have yet to exhibit any *overall* superiority over other media and techniques. On many aspects of the use of gaming-simulation, planners are still exceedingly ignorant. At present, as this study has attempted to reveal, little is known about the learning opportunities presented, the rate the student acquires this learning, the method by which he acquires it and at what cost. These are but some of the immediate problems which have to be faced. Ultimately, more reliable information is needed on these subjects just as more must be known about the nature and juxtaposition of both old and new methods alike. Obviously, this last point covers an area where the concept of a total instructional simulation system is an important one and it is an area where more knowledgeable appreciation is needed if all facets of gaming procedures are to be evaluated dispassionately. In brief, greater efforts are required to define how gaming activities can best be used and how best they can be married to other techniques and fully integrated within the curriculum.

Finally, with regard to the general receptiveness to educational change at university level, it might be said that something is lacking. The innovations and vitality brought to infant and junior schools by, amongst others, the Schools Council, the Nuffield Foundation and numerous individual teachers should be equally present in professional education. Responsibilities at this academic level must be more widely distributed. The concerted and broadly based support and involvement of professional institutions, research centres, universities and a diversity of practitioners is obviously vital to current needs. It is not unreasonable to expect all these parties to consult together more regularly and more profitably and, where necessary, to co-operate and/or complement each other. The evolution and strength of planning education rests on many shoulders and it remains to be seen to what extent this responsibility is accepted.

CONCLUDING REMARKS

In summary, urban and regional gaming simulations are still both very much of a novelty and an unknown quantity. There is a lack of design and operational experience and an absence of objective evaluation studies. To date, research effort has been concentrated on the evolution of experimental prototypes and, as yet, there are very few clear indications of the specific usefulness of the technique.

At this point, it is appropriate to briefly restate what urban development games are possibly attempting to accomplish at a higher educational level. It seems clear that the instructional simulations described here *do not* demonstrate what will happen in a set situation; at best they illustrate what *could* happen and perhaps, more importantly, reveal something of how the process of change occurs. The game is a means of exercising, extending and hopefully improving the participant's knowledge, skill and understanding. The player's comprehension of a dynamic situation and his ability to manipulate and manage the simulated system is put to the test. In exploiting the situation as he sees it he is free to: interact with other decision-makers; experiment with alternative strategies; prove his worth as best he can; and, at the same time, witness and become part of the cybernetic process of *learning to learn.*

Because games tend to emphasize the provision of insights into processes rather than the communication of distinct facts or specific skills, it is often very difficult to specify their pedagogic function and content, and hence equally difficult to validate any particular model. The impressionistic nature of many innovators' reports is apparently fully appreciated by the authors concerned and efforts are being made to ameliorate this situation.

Thus it has been concluded that the case for or against urban development gaming stands largely unproven, and present indications suggest that this situation will not easily be resolved. However, the large body of scholars who have devoted time to work in this field and the sheer weight of favourable testimonies is itself encouraging, not to mention the fact that the reports emanating from such sources have been, as yet, in no way convincingly challenged or disputed. At the same time, no evidence has been presented to suggest that gaming techniques in their present form can supplant older established methods. Above all, it has been pointed out that the progress achieved in evaluating gaming procedures, in a relatively short period, although limited, is in no way overshadowed by the results of attention devoted to similar procedures in other fields *or* by the findings of research into other teaching methods at university level. In fact, there is good reason to believe that, in planning education, the assessments of gaming-simulation have already progressed further than appraisals made in connection with any other teaching techniques used in this area. This is not to say that much more work and increasingly precise validatory data is not urgently needed; indeed, until this is forthcoming, many people, quite rightly, will probably find it difficult to accept gaming-simulations as little more than a rational and provocative means for structuring and manipulating discussion.

This publication has thus served to further underline the need for a far more positive commitment to planning education's aims, methods and procedures. On the assumption that ways of improving current methods and procedures can always be found, it seems vital for all concerned with the pedagogic process to be more active in promoting, utilizing and evaluating instruction innovations and experiments. It is from this broadly based involvement that greater progress in education may well be accomplished.

In the case of urban development studies, as yet there is little evidence of attention being devoted to instructional methods on the lines fostered, for example, by the school system pioneers. This publication has sought, in this respect, to encourage: a greater interest in the continuous re-examination of instructional systems; and a more enlightened attitude towards educational experimentation. It is certainly not the intention of the writer to substitute untried opinion and predisposition for time-honoured attitudes; instead, the object of this work was to examine *one* possible source of educational improvement which, from the outset, appeared worthy of greater interest, wider understanding and hence increased development and trial.

Thus, the burden of this volume has been concerned with the evolution of a possible approach to urban studies which was seen as a promising means to engage, motivate and stimulate students beyond the usual levels of commitment. The book is neither a new theoretical statement nor a complete operational guide. Rather it is an attempt to provide through a series of perspectives a framework by which the reader may start to build up a fuller awareness of a technique. The inter-relationship of various facets of gaming-simulation systems have been shown without claiming that each dimension encountered interlocks with others or comprises a distinct part of a clearly identifiable totality. In short more questions have been raised than perhaps answered and certainly much more needs to be known if rational conclusions are to be reached. However, it is hoped that some of the issues raised have sufficient relevance to encourage others to undertake additional studies towards discovering educational improvements which will be more fruitful in the future. If the impact of instructional simulation does no more than set in motion and encourage greater objective assessment in planning education it will have produced a worthwhile result.

At this point in time and on the basis of what is now known, some of which has been reported here, the writer has no hesitation in stating that it seems highly likely that gaming procedures will continue to multiply and, in all probability, be more widely and efficiently used. Many of today's urban development games will possibly be superseded and more sophisticated models appropriate to specific settings seem likely to emerge. The technology supporting such development shows every sign of continuing to improve as does the discrimination of the user. Ultimately, it seems that a greater emphasis will be placed on the production and dissemination of total learning systems.

From the foregoing, it should be evident that, despite the considerable doubt surrounding planners' approaches to urban and regional system simulation in higher education, the writer adheres to his continuing belief that the technique has some instructional relevance and considerable potential. However, the reader should take note that this sense of conviction has not been reached without reservation and it bears repetition that experience to date is *limited* and the place occupied by the technique is *exceedingly modest* and in few ways well integrated into overall urban studies programmes. In short, the technique is not without difficulties and, as is usual in such studies, it is suggested that to be aware of these problems is the important first step towards solving them.

Finally, it is hoped that some of the perspectives presented may have established the groundwork from which more informed examination and discussion can take place. In particular, by revealing what has been achieved by, and what is available to, urban specialists, perhaps more systematic investigations and rigorous trials will be stimulated or provoked throughout the entire educational system. The gulf between a series of unrelated teaching innovations and widespread academic usage must still be narrowed but it is believed that this work has shown that, within a decade, the gap has already been considerably lessened and that knowledge of the technique is no longer restricted to relatively small groups of isolated academics, as evident 10 years ago. In sum, an encouraging stage has been reached, but this modest achievement still leaves much to be done if this particular approach to the urban development process is to be continuously refined and properly evaluated. However, if simulation systems based on gaming procedures were to realize even a small part of their instructional promise, this effort would appear to be well justified and a similar book in a further decade would obviously have much to report.

Appendix 1

Notes and operating instructions on a land use gaming-simulation system

This appendix endeavours to comment briefly on the design and use of a prototype instructional simulation system for readers wishing to participate in a simple urban development gaming exercise. Little simulation knowledge or gaming experience is assumed and it is emphasized that these notes are intended to provide only an introduction to *one* form of operational gaming procedure currently being used experimentally in urban land use planning studies.

The model to be described was designed by the author in association with R. N. Maddison, formerly of the University of Sheffield's Computer Laboratory. In addition to Dr Maddison's continual guidance the writer is especially grateful to all of those who have contributed in some way to the development of this model. In particular, a considerable debt is owed to A. G. Feldt, R. D. Duke, K. R. Carter and R. Slevin for their inspiration and wise counsel during the design process, associated experimental trials and the formulation of material for this appendix.

The appendix falls into three major parts. The first part explains the origins and development of a prototype LAND USE GAMING-SIMULATION (LUGS) and specific reference is made to certain design and operational features. The second part describes the terminology, equipment and roles involved in the basic model and the third and final part outlines preliminary stages in the standard operating procedure.

A. INTRODUCTION

Encouraged by the work of Hendricks (1960), Duke (1964) and Feldt (1965), it was decided early in 1966 to subject an operational gaming-simulation to a series of trials at the University of Sheffield. From the limited range of planning simulations available, for reasons of simplicity, the CORNELL LAND USE GAME became the experimental model. This introduction brought about a desire to build a simulation more appropriate to local conditions, in terms of both statistical data and educational needs (see Taylor and Maddison 1968).

The resulting LAND USE GAMING-SIMULATION (LUGS) is a derivative of the basic CLUG model and a considerable debt is owed to Allan Feldt for generously making so much of his own material available for unrestricted use. Sheffield's LUGS model, like the Cornell Game, concentrates on the urban growth process in terms of selected and well-defined characteristics. Any number of teams can take part and can exercise certain entrepreneurial and democratic rights as and when financial and political circumstances allow. All team decisions are freely selected from a defined range of choice and the game is built up of sequential decisions concerning either private speculation or community development. The game can be terminated at any point and no optimum strategy is prescribed.

Original objectives

The LAND USE GAMING-SIMULATION was designed in an effort: (1) to create an environment for studying dynamic urban situations or processes which defy *economical* description in literary or mathematical terms; (2) to create a rudimentary means of integrating urban relationships involving a wide range of tangible and intangible variables; (3) to create an 'open-ended' educational framework which would allow maximum opportunity for incorporating a variety of elements relevant to differing educational objectives and teaching situations; (4) to develop a simple introduction to the design and use of gaming-simulations with specific reference to the study of urban planning.

In setting up these tentative goals, certain constraints were identified. The proposed gaming-simulation was required for operation: (*a*) at pre-university and early undergraduate level; (*b*) with differing sizes of study groups not in excess of 25 in number; (*c*) with student planners and potential architects, engineers, economists, geographers and sociologists; (*d*) with participants possessing a minimum of mathematical ability; (*e*) without the use of a computer; (*f*) in relatively short teaching sessions, i.e. with approximately 2–3 hours simulating 10–15 years.

Features of the operational game

To the novice, the game, devised to meet the preceding requirements has many of the strategic formalities found in chess and calls for mock performances normally associated with educational role-playing. The game simulates the growth of hypothetical settlements in terms of certain well-defined characteristics (e.g. land use classifications and economic returns) and within the limits of specified rules or constraints (e.g. set geographic zones and constant work force relationships). Each team provides and/or employs labour as economic circumstances demand. The Local Authority team endeavours to control 'play' and serves as the administrative and organizational mechanism behind the model. Decision outcomes are deterministic; thus by reference to tables or formulae, results are explicitly assigned to specific decisions. A single stochastic, or random element (contingencies representative of floods, gales and fires) is introduced as a token reminder of the hazards of partial control which spring from natural as well as human variables. Profitable economic participation largely depends upon a full appreciation of the relationships amongst the variables and upon an ability to predict the behaviour of fellow decision-makers.

It has been stated that any number of teams can ' play ' the ' game ' but administrative complexities encountered with large numbers suggest that from three to five teams, of one to three members per team, should be actively engaged in demonstration sessions. The local government team requires one or preferably two persons to represent the local authority role and to act as administrators. They are expected to explain the game and individual steps whilst ensuring that the simulation exercise progresses at a reasonable speed. In their capacity as a local planning authority they are supported by at least one player acting as the professional planning team. All other teams play the combined roles of the real estate speculators, the labour force and the elected local authority representatives. Of the members of a team, one normally does the book-keeping and the other(s) make decisions and note detailed progress. Once a degree of proficiency is displayed changes in roles and duties are encouraged and players are invited to proceed with elaborations of the game which introduce separate teams of politicians, financiers and the public.

The physical model, which displays the form of the simulated community, is built upon a subdivided board composed of a hundred 50 mm (2 in.) squares. Each square represents a plot of land 0.5 km (¼ mile) square, i.e. 25 hectares (40 acres). Between grid squares there are three 5 mm (¼ in.) spaces for indicating the presence, if any, of (*a*) communication routes, called ' links ' and representing *A* or *B* class roads and associated facilities such as gas, electricity, water and sewerage; and (*b*) land use or ' zoning ' restrictions.

Before and during the game, areas may be ' zoned ' or designated for specific uses. This zoning, if any, is indicated by lines drawn on the perimeter of each plot. Coloured tiles are used as symbols, or pieces, representing the basic categories of land use, e.g. industrial, commercial, public service, residential. The size of each piece corresponds to the number of employees in residence or actually employed. For reasons of simplicity transit terminals, public services and open spaces are not regarded as financial employers within the model, but an appropriate residental block must be accommodated on the grid reference sheet when public services are constructed.

The game proceeds in periods, each representing two years of real time, which are numbered multiples of two. For every period each team uses a single duplicated sheet for recording calculations. This period record sheet (*Form* 10) is completed by each team and gives short explanations of the sequence of operations to be undertaken and the requisite space to enter appropriate details. On average only about 10–15 numbers need to be entered by each team per period.

Each team is given specific resources at the commencement of the game. These resources are distributed in the form of operating assets and a cash balance. The exact nature of each team's allocation is dependent upon: the educational objectives involved; the number of teams employed; the participants' experience and familiarity with the model; and the initial form of the simulated environment.

The speed of play can range from one to two periods per hour, as modulated by the administrator. To ensure, for example, two periods per hour, if after 20 minutes the current period is not drawing to a close, then it is prematurely terminated; no further decisions in that period are allowed, the necessary book-keeping is immediately completed and the next period is commenced. This forced termination should occur only rarely, when novice participants are involved.

The steps printed on each period record sheet are self explanatory and are given in the order in which they are normally encountered. It remains to explain that, where voting occurs, 'approval' is by a simple majority on a show of hands. Each team has one vote and the administrator, representing the local authority has, when necessary, a casting vote. Voting issues are usually concerned with straight choices, e.g. for, or against, a single proposal. Consequently a separate vote is taken on each separate issue as, and when, required.

B. EQUIPMENT, ROLES AND TERMINOLOGY

The full equipment comprises: a grid reference sheet (see *Form* 2) which is mounted under a protective acetate sheet, and over a flat magnetic metal display base; a set of plastic-faced magnetic tiles, shaped and coloured as set out in *Form* 3 and representing various units in the model; chinagraph marking pencils for use on the acetate sheet and the plastic faces of the magnetic tiles (as this will wipe off these surfaces easily), two dice and a supply of forms – the data sheets, the period record sheets and rates record sheets for the participants. It is an asset to have an electronic calculator available, but this is not essential.

Grid reference sheet

The grid reference sheet acts as the display framework for the physical model; and the size and scale of the grid are determined by the exercise to be undertaken. A relatively modest grid of 100 equal 50 mm (2 in.) squares is generally considered adequate for introductory purposes and this is shown in a scale plan as *Form* 2. Each square represents 0.5 km (¼ mile) square or roughly the equivalent of 25 hectares (40 acres) – hence the overall scale may be considered as approximately 1 : 10,000.

Teams and players

The teams and players are the motivating force behind the model. Any number of players can be involved but the administrative difficulties which are encountered with unlimited participants suggest that players should be grouped into teams of from two to five members. During the initiation stages it is recommended that only three entrepreneurial teams, of two members per team, are actively engaged. In addition there is the administrator of the game and an individual or group of people to act as professional planners. Other elaborations can include teams to represent the general public, elected politicians, and independent financiers, but these are not considered in this outline description of the basic introductory version of the LUGS model (see *Form* 12).

Timing

Each period of play represents approximately two years duration; hence in referring to periods, the first is known as period 0, the second as period 2, the third as period 4, and so on, using even numbers only.

It is considered realistic to limit the actual time taken for the computation of all operations in each period and, initially, a maximum of 40 minutes per period may be required; but subsequently 25 minutes per period should be possible as a standard operating time. The length of each exercise, in terms of periods played (and hence real time portrayed), is flexible and need not be publicly defined before commencing proceedings. It is hoped that this element of secrecy encourages continuous optimization of resources and strengthens the realization of planning as a continuous process! It is stressed, however, that each exercise can be designed and run to a predetermined schedule. To date, experience has shown that ten to twelve 'two-year' periods (i.e. three to five hours playing time) might be regarded as the minimum operating time.

Finance

All transactions within the game are conducted on a monetary basis, and they are recorded in units which are equivalent to £1,000 in real life: hence 30 units = £30,000.

Each entrepreneurial team is allocated a specific capital sum at the commencement of the game; the actual amount allocated is flexible and is entirely dependent upon the objectives of the particular exercise undertaken. The amount given to each entrepreneurial team need not be publicly disclosed and this sum may vary from team to team. It is suggested that 5,000 units (i.e. £5,000,000) is a reasonable initial sum for a novice team with operating assets of approximately 20,000 units (i.e. £20,000,000). If a new community is to be developed on a virgin site (i.e. a blank grid), then a minimum of 15,000 units is recommended when, for example, three teams of novices are playing.

Operating assets

In addition to the capital with which each team commences the game (which can be converted into capital assets in the form of real estate land and building blocks – and from which income is derived), a team may start with operating assets'. This team covers the income-earning land-use activities (i.e. industrial, commercial and residential) and, as in the case of capital assets, initial operating assets may be allocated in any manner appropriate to the purposes of the teaching exercise.

Net income

For each trading investment (i.e. industrial, commercial or residential) there is a related net trading balance. This balance is the product of a net income formula shown on *Form* 9 (Income from activity blocks). It will be noted that the variable in each formula relates to the transportation factor. With industrial and commercial users, the transportation factor is the distance between the user and the *nearest* transit block. For residential activity blocks, the transportation factor is the distance between the residential block and its designated place of employment. In both cases, distances are calculated by counting the number of grid junctions encountered along the shortest route between the mid-points of links adjoining the plots concerned. Hence the degree of accessibility between certain land uses and communication centres, and also between places of residence and employment, is a fundamental determinant of successful economic participation in the game. Consequently, most land-use decisions will be determined largely by current or anticipated patterns of land-use development.

Residential activity blocks are economically related to employing units (two residential blocks (RB's) to one industrial or commercial block) and hence revenue only accrues from any block if its complementary blocks are present. It is not essential that complementary blocks be owned by the same entrepreneurial team; hence, for example, one team may build residential blocks for another team's industrial block, and so on.

Administrator

The administrator plays the role of the local authority, in administering the game and provides the organizational mechanism behind the gaming-simulation exercise. Ultimate control rests with the local authority, and its specific functions may be listed as:

1. Instructor of operating procedures and arbitrator in all disputes.
2. Recorder for the grid reference sheet and auditor of period record sheets.
3. Loan facility operator.
4. Auctioneer for all land transactions.
5. Assessor of rates allocation.
6. Assessor of contingency costs.

The above functions could be performed reasonably by a single person but in the interests of speed and general efficiency it is an advantage for two people to be concerned with this role when more than ten participants are involved.

The planners

Because the administrator is essentially an impartial operator of the game, the role of the planners is taken by a separate person or team of people. Their major task is to prepare a development plan for the model and to modify and amend this as necessary during the course of the game.

The public, the politicians and the financiers

In this simplified version of the model, all community decisions are made by the entrepreneurial teams. However, in a more elaborate development of the model it is possible to add teams or individuals representing the public and the politicians within the decision-making framework. In addition, a team can be added to represent the financiers who are able to offer additional loan facilities for short-term investments at higher rates of interest and at different stages of the game from those offered by the administrator.

The development plan

These proposals originate from the planners and are presented to the community by the administrator who arbitrates in the approval of the contents of the plan, which must be made by a majority vote of the participating entrepreneurial teams. The plan may be revised or discussed, and modifications made at the set stage during any period of play. The development plan may designate the following land use allocations (delineated by colour):

Industry	—red
Commercial	—blue
Residential	—yellow
Public buildings	—black dots on a white ground
Public open space	—green
Transit blocks	—black
Communication links	—black
'White land'	—white

Programming can take the form of restrictions on the availability of land, restrictions on the amount of building permitted, including building in the local authority sector, and restrictions on the building of communication links or transit blocks – such restrictions may be imposed on the community without its approval, as if it were from the central government agency.

Community decisions

Several decisions (including the development plan approval) in the course of the game have to be made by the community as a whole and, in this simplified version of the model, this decision-making process is conducted by a majority vote from the participating entrepreneurial teams. This, of course, means that the decisions are influenced by vested interests; consequently in more elaborate versions of the game the public and the politicians are brought into the decision-taking process as additional participating teams with appropriate voting rights.

C. OPERATIONAL PROCEDURE

During the exercise, each team must consider and decide whether to act on any issues listed in a set sequence of operations described briefly on the period record sheet. Each of these actions and their resultant effects are jointly recorded on the grid reference sheet and on individual team period record sheets.

The grid reference sheet

This document has already been described and during the game it is managed and controlled by the local authority (i.e. the administrator). Upon it are shown the following features:

1. Existing physical restrictions.
2. Development plan proposals for land use, communications, programming, etc.
3. Land ownership and the price paid for the private plots (see *Form* 7).
4. Physical constructions on the site: transit blocks, industrial blocks, commercial blocks, public building blocks, public open space, residential blocks and communication links (see *Form* 3 for details).

Each industrial and commercial block displays its team ownership and the period in which it was constructed. So too do residential blocks on which, in addition, is marked the grid reference of the assigned place of work. Local authority housing (i.e. assigned to public building blocks) carry only the grid reference of their work places. This documentation is fully explained on *Form* 7.

Only *one* activity block may be constructed on any *one* plot unless the site is specifically designated otherwise.

The period record sheet

Each entrepreneurial team completes an individual period record sheet for each period of play. This sheet contains a detailed financial breakdown for the relevant transactions of the team. The recording of these transactions is in three columns; debit, credit, and balance. It is easier to record each transaction separately in the debit or credit columns as appropriate, and then to total each of these in order to adjust the balance column at the end of the period. During the course of any period, a team's balance may fall to a negative value, but *only* if it is certain that by the end of that period's transactions (i.e. stage 17 of the period record sheet) a positive balance will have been restored.

Individual stages on the period record sheet (see *Form* 10) are now described in the sequence in which they are *normally* encountered.

Stage 1 : Enter cash balance at start of each period. After receiving an initial sum of capital at the commencement of the play, this operation should become a mechanical piece of administration. It merely requires the record keeper of each team to transfer the balance at the end of the previous period (stage 17) into the balance column at the start of the current period.

Stage 2 : Receive interest on cash balance. At the commencement of each new period, interest is calculated on the balance held by each team at the start of the *previous* period. As a period is the equivalent of two years in real time, a 10% interest rate is used as an approximation for the compound interest accumulated over this two-year time span.

Stage 3 : Consider development plan proposals as appropriate. At this stage in the proceedings any team may raise any development plan considerations for open discussion, and community decision.

Stage 4 : Negotiate loans as required. The local authority (i.e. the administrator) has the ability to loan cash in multiples of 4,000 units (i.e. £4,000,000) and repayment (with interest) is carried over the five subsequent periods. If, for example, 4,000 units are borrowed, then in the five subsequent periods, *five* separate 1,000 unit instalments must be repaid. Loans may *only* be negotiated on *periods 10, 20, 30, 40,* etc., *and* when a sum not in excess of four times the previous period's total income is required.

Stage 5 : Construct new communication links as required. Communication links represent a number of services normally associated with road development and community expansion. These services include such items as sewerage, electricity, water, gas, telephones and so on. Their construction is a community decision within the confines of the development plan and the grid reference framework. Those activity blocks represented by *squares* on the grid reference sheet (see *Form* 3) require at least *two* communication links adjacent to the plot before construction is permitted; those activity blocks represented by triangles or rectangles only require *one* communication link adjacent to them. In both cases the required links must be connected with the overall communication system on the grid. There is no capital cost to any team for the construction of communication links but each link does add 30 units (i.e. £30,000) to the total rates assessment per period.

Stage 6 : Construct new public building blocks. Residential blocks require the support of public services such as schools, health and welfare, libraries and so on; these are represented in the model by the public building block, the building of which is the subject of another community decision. Such a decision cannot over-ride the provisions of the approved development plan. The need for public buildings to be related to the residential blocks is explained in stage 10 (commuter deductions) and on *Form* 6. There is no capital cost to any team for the construction of public building blocks but each block adds 30 units to the total rates assessment per period.

One residential block must be constructed by the community (at no capital cost) for each public buildings block, on a separate plot of land owned by the local authority. This unit *must* be fully serviced – i.e. there must be no commuter deductions for the plot on which the residences are erected.

Stage 7 : Construct public open space (POS) as required. As with public buildings, there is a need to provide public open space for active and passive recreation within the vicinity of residential blocks and this provision is met by the public open space block. Again it is a community decision to provide this, and it must fall within the provisions of the development plan proposals. There is no capital cost to any team but each POS block adds a further 30 units to the total rates assessment per period.

Stage 8 : Construct new transit blocks as required. The transit block represents the combination of transportation terminal facilities at the community level; and the economic viability of industrial and commercial blocks is determined by the locational relationship between these land uses and the transit block. Once again the construction of a transit block requires a community decision by the participating teams, within the provisions of the development plan. Its cost is again in the form of an addition to the total rates assessment, adding 300 units (i.e. £300,000) per block constructed per period.

Stage 9 : Calculate the cost of contingencies. On periods 4, 8, 12, 16, etc., the local authority determines the cost of contingencies such as gales, fires or floods. Each team throws two dice and, by so doing, the local authority determines the team's actual losses with reference to *Form* 5. This chart relates to the total age of buildings in the team's ownership, the risk of loss and the losing dice numbers, to the sum forfeited as a contingency penalty. The penalty represents the loss to the team over and above that for which it is insured and appears as a cash debit on the period record sheet.

Stage 10 : Commuter deductions as applicable. As mentioned in stages 5, 6 and 7, each residential block requires certain facilities within easy reach – namely commercial, public building and public open space blocks. Commuter deductions are penalties which are incurred by each residential block which does not have one of *each* of these facilities in any of its eight adjoining plots. *Form* 6 demonstrates this. For each facility that is lacking, a deduction of 50 units (i.e. £50,000) per plot is made; and if any facility does not appear on the grid at all, then this deduction is increased to 100 units per plot per absent facility. In the case of residential blocks related by employment to public building blocks these can *only* be built on plots for which there would be *no* commuter deduction.

Stage 11 : Buy or sell land. At this stage in the period the local authority auctions publicly owned plots. Bids are accepted in units of 100 (i.e. £100,000) and the highest bid secures the plot. Individual teams may also auction their private real estate assets. It must be stressed that the price of land is not fixed; but the vendor may set a reserve price for its sale. On the grid reference sheet the team purchasing the land and the price paid for it is recorded (see *Form* 7); whilst on the period record sheet the team making the purchase records the grid reference of the plot, and the price paid or received for each individual transaction. When all transactions are complete, the total debits and credits from this stage are transferred to the running-total columns for debit and credit on the right of the sheet.

Stage 12 : Construct buildings. Each team may choose to construct certain types of building blocks, namely industrial and commercial and the residential blocks associated with either of these employment activities. The price paid for the blocks is fixed according to a predetermined rate which reduces as the number of identical blocks built in any *one* period increases. Full details of these costs are given on *Form* 8. It will be noticed from this table that reductions of up to 30% are obtainable by specialization in one facility. On the period record sheet each team records details of the grid reference of plots developed, the type of block constructed and its cost. While on the grid reference sheet, the magnetic tile symbol for the block is placed on the appropriate plot by the local authority, and on it is recorded the period in which it was constructed (in the case of residential blocks the grid reference of the assigned employing block) (see *Form* 7). When all of the building operations are finished, then the total cost is transferred into the running-total debit column on the right-hand side of the period record sheet.

Stage 13 : Net income from new buildings. The relevant net income formula (see *Form* 9) must be computed for every *new* building constructed in the preceding stage 12. To complete the formula, the total communication link distance must be known (the transportation factor), and from this a simple arithmetical calculation will give the income. The measurement of distances is taken from the centre points of the communication links along the shortest route to the transit or employing block (as appropriate); when this distance is directly across a communication link (i.e. between two adjoining plots separated by a communication link), then no deduction is made. *No* income is received from any industrial block or commercial block until *two* related residential blocks are provided and, conversely, no income accrues from residential blocks for which there is no employing block.

Stage 14 : Continuing net income from existing buildings. Under normal circumstances this figure can be obtained from stage 15 in the previous period. However, if a team has engaged in any demolition or has been affected by additions to the communications network, then amendments in the total figure will be necessary, subtracting the loss or gain in the recurring income from that which appeared in stage 15 on the previous period's record sheet.

Stage 15 : New total net income from all buildings. This figure can be obtained by adding the totals from items (13) and (14) above. It can then be transferred into the running credit column on the period record sheet; it will also be of value in the calculation of stage 14 in the subsequent period.

Stage 16 : Calculate rates assessment. It is necessary for the community to bear the cost of all the community-controlled facilities, i.e. communication links, transit blocks, public buildings, public open space; hence the local authority calculates the cost of these facilities by a set formula and then levies an appropriate proportion of this total cost on each privately owned plot. This calculation is clearly shown on the rates record sheet (*Form* 11) which is used by the local authority to record the transactions during each period. Each item forfeits rates for each plot on which a block has been constructed and which they own; and if any team defaults on their ability to pay this rate levy, then they are required to sell enough holding to meet their commitments.

Stage 17 : Record the balance at the end of the period. The period ends with all of the teams compiling their own assessment of the results of their decisions. The debit and credit columns are totalled and the original balance for the period is adjusted accordingly. It is now possible to commence the next period, the figure resulting from this stage being carried forward to stage 1 on the new record sheet.

FORM 1. LIST OF BASIC OPERATING EQUIPMENT

Supplies of: Introductory notes (*Form* 4)
Table of contingency costs (*Form* 5)
Table of costs of building blocks (*Form* 8)
Table of income from activity blocks (*Form* 9)
Period record sheets (*Form* 10)
Rates record sheets (*Form* 11)

1 grid reference sheet (*Form* 2)
1 transparent acetate sheet to cover grid reference sheet
1 magnetic metal display base to go under grid reference sheet (only required if magnetic tiles are used)
1 black chinagraph pencil, or equivalent
1 white chinagraph pencil, or equivalent
2 dice

Sets of coloured markers or magnetic tiles as shown in *Form* 3: although magnetic tiles are ideal, they may be replaced by other suitably shaped and coloured material such as card or wood.

It should be stressed that the preceding list covers the minimum amount of operating equipment necessary to mount an introductory gaming-simulation exercise using the LUGS model. Additional materials are required, however, if the model is to be used other than as a preliminary demonstration of a simulation technique.

FORM 2. THE GRID REFERENCE SHEET

	A	B	C	D	E	F	G	H	J	K
1										
2										
3										
4										
5										
6										
7										
8										
9										
10										

The main grid reference sheet is shown above in plan form. Each of the referenced plots is 50 mm. × 50 mm., and three 5 mm. wide bands separate the plots (or 2″ × 2″ separated by three $\frac{1}{4}$″ wide bands). The central band accommodates the communication links; the others are for zoning delineation. Hence the overall grid reference sheet is a square approximately 655 mm. (28″) square. Each plot represents a square of 0.5 km. sides—thus the whole sheet represents an area of approximately 25 sq. km. ($6\frac{1}{4}$ sq. miles).

FORM 3. THE SYMBOLS USED FOR BUILDING BLOCKS IN THE MODEL

Symbol	Real world description	Initial cost per block (see Form 8)	No. of supporting residential blocks required in model	No. of adjacent communications links required in model
INDUSTRIAL BLOCK				
red	An amalgam of various types of industrial land uses, ranging from heavy through to light industry	5000	2	2
COMMERCIAL BLOCK				
blue	Commercial and local centre of town including shops, offices and services, banks, building societies, garages, etc.	2500	2	2
TRANSIT BLOCK				
black	The road, rail, air and water transport centre for goods and passengers passing into or out of the area	*	0	2
PUBLIC BUILDINGS BLOCK				
white	The local authority offices, primary and secondary schools, libraries, health and welfare facilities, etc.	*	1	1
PUBLIC OPEN SPACE				
green	All public open space for active recreation (sports grounds, etc.) and for sitting out (parkland and commonland)	*	0	1
RESIDENTIAL BLOCK				
yellow	All forms of dwelling unit, in low-rise houses and bungalows, high-rise flats and maisonettes, etc.	1000	—	1
COMMUNICATION LINK				
black	All services to land-use units, i.e.: roads and paths, sewers and refuse disposal, water, electricity, gas, post, etc.	*	—	—

* indicates that these blocks do not impose direct private costs on the entrepreneurial teams, but impose additional costs on the rates assessment funds (see *Form* 11).

FORM 4. BRIEFING NOTES ON AN INTRODUCTORY VERSION OF THE LUGS MODEL

GENERAL DESCRIPTION OF THE GAME

To the novice the game has many strategic formalities used in chess and calls for mock performances normally associated with educational role-playing. The game simulates the growth of hypothetical settlements in terms of certain well-defined characteristics (e.g. land-use categories and economic returns) and within the limits of specified rules or constraints (e.g. static geographic zones and constant work-force relationships). Each team provides and/or employs labour and is involved in community decision-making processes. Administrative and organizational procedures are carried out by the local authority team.

EQUIPMENT AND PROCEDURE

A board is used as a physical model and represents an area of land to be developed. The board is divided into quarter-mile squares, with a space between the squares for the siting of roads. Each plot is given a grid reference letter and number for location purposes. Coloured tiles are placed on the board to indicate the following activities:

Red square	One industrial block	2 RB's employed
Blue square	One commercial block	2 RB's employed
White triangle	One public building block	1 RB employed
Green triangle	One public open space block	1 RB employed
Black square	One transit block	2 RB's employed
Yellow rectangle	One residential block (RB)	
Black strip	One communication link	

Each entrepreneurial team compiles a single period record sheet for each period of the game; one period represents two years of real life. The periods are numbered by consecutive *even* numbers, commencing at 0. The record sheet gives short explanations of all operations in the order in which they are encountered in a full cycle. At the commencement of play, the capital amount allocated to each team is entered under the appropriate column on their period record sheet. As resources are utilized in the purchase of land, construction of buildings, etc., the relevant financial adjustments are made according to the prescribed allotments.

Communication links are not owned by any one team, but by the community in general. The same is true of public buildings blocks, public open space blocks and transit blocks. A majority vote is needed for the construction of these items. Each new link, public building block, POS, or transit block adds specific amounts to the total rates.

Nothing may be constructed on a plot which lacks a communication link on one of its sides. For industry, commerce and transit blocks, links are needed on at least two sides of the block. For the others, at least one link must run adjacent to the plot. Also, before any non-residence block can be built, it must be ascertained whether sufficient residential blocks exist to house the additional workers. These residences may be owned either by the team employing the residents or by another team with whom an arrangement has been made. Only *one* land use may be located on one plot unless otherwise permitted by the community's development plan.

Public building blocks, and POS blocks engage employees from one RB, which must be built on publicly owned land on the grid; transit blocks employ 2 RB's, which must be similarly accommodated.

Distances are calculated along the communication links which exist when the buildings are first erected; but if new links are built in later rounds, the distances previously calculated may be changed. A distance is calculated by counting the number of grid-junctions encountered along the shortest route between the mid-points of the links adjacent to the plots concerned.

GAME STRATEGY

The game may continue for any prescribed period and when play is concluded the 'pay-offs' or decision outcomes will be discussed in terms of performance indices.

Performance and in some cases survival is dependent upon the outcomes generated by the model. Successful participation depends on a full appreciation of the relationships amongst the variables and upon an ability to predict the behaviour of other players.

FORM 5. CONTINGENCY COSTS (STAGE 9)

Total age of all buildings owned by team	Losing dice numbers	Probability of loss	Contingency penalty
2 4 6	2 7 9	0.306	2,200 2,400 2,600
8 10 12	3 7 8	0.361	2,800 3,000 3,200
14 16 18	4 7 8	0.417	3,400 3,600 3,800
20 22 24	6 7 8	0.445	4,000 4,200 4,400
26 28 30	3 6 7 8	0.500	4,600 4,800 5,000
32 34 36	5 6 7 8	0.556	5,200 5,400 5,600
38 40 42	4 6 7 8 10	0.611	5,800 6,000 6,200
44 46 48	5 6 7 8 9	0.667	6,400 6,600 6,800
50 52 54	2 5 6 7 8 9	0.695	7,000 7,200 7,400
56 58 60 and above	4 5 6 7 8 9	0.750	7,600 7,800 8,000

The appropriate contingency penalty is determined during stage 9 on the designated periods 4, 8, 12, 16, etc.

FORM 6. COMMUTER DEDUCTIONS (STAGE 10)

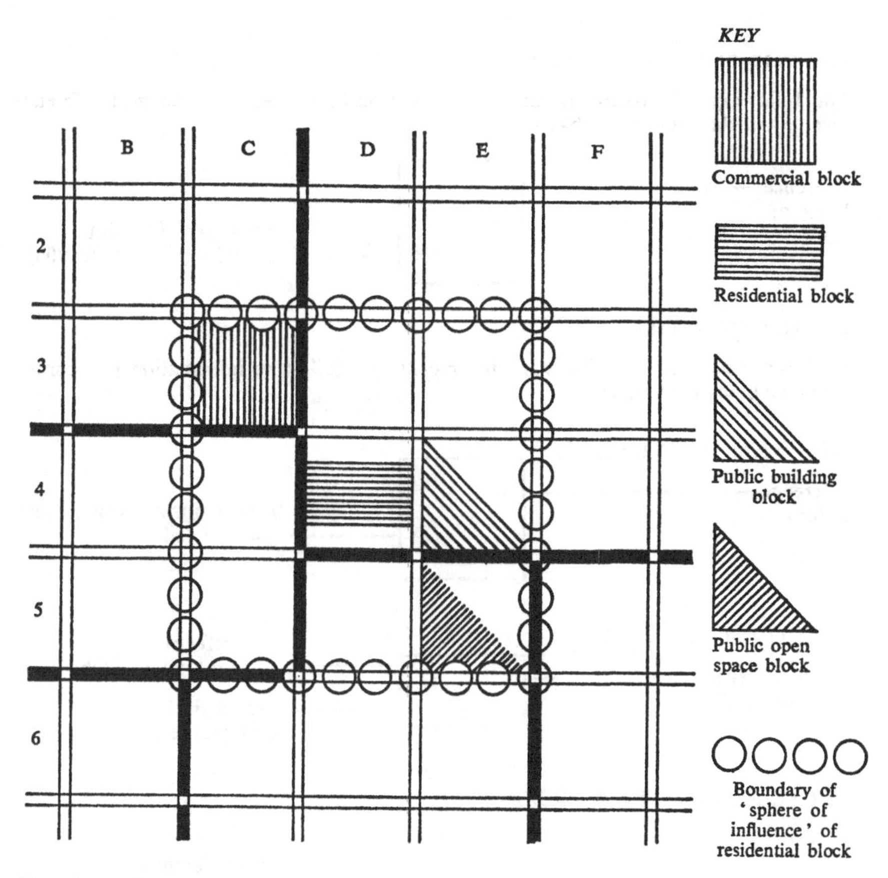

For the residential block shown above in square D4 there will be no commuter deductions, because within the eight surrounding squares (shown in the notation as the 'sphere of influence' of the residential block) there are:

1 commercial block in square C3
1 public buildings block in square E4
1 public open space block in square E5

However, if any one of these facilities were not located within the sphere of influence, then commuter deductions would be made for each absent facility

if the facility exists elsewhere on the playing grid, @ 50 units per residential block
if the facility is not present on any part of the grid, @ 100 units per residential block

Housing built by the local authority related to a public buildings block may only be constructed on plots for which no commuter deductions are appropriate.

FORM 7. INFORMATION RECORDED ON GRID REFERENCE SHEET AND MAGNETIC TILES

THE GRID REFERENCE SHEET

The following information is recorded on the acetate sheet over the grid reference sheet when the plot is purchased:

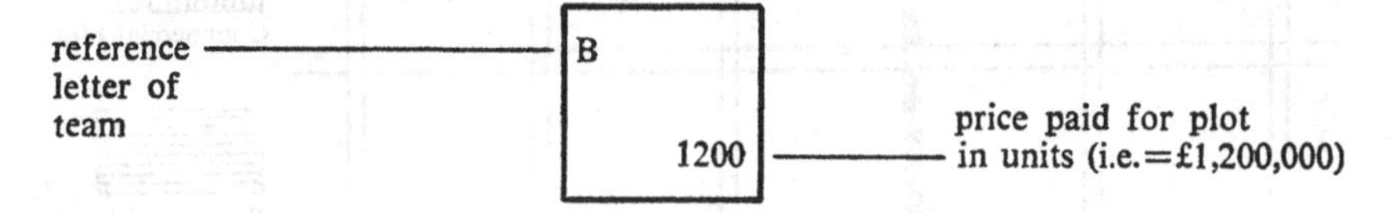

THE MAGNETIC TILES

On the magnetic tiles, which mark the blocks, the following information is recorded when the block is erected:

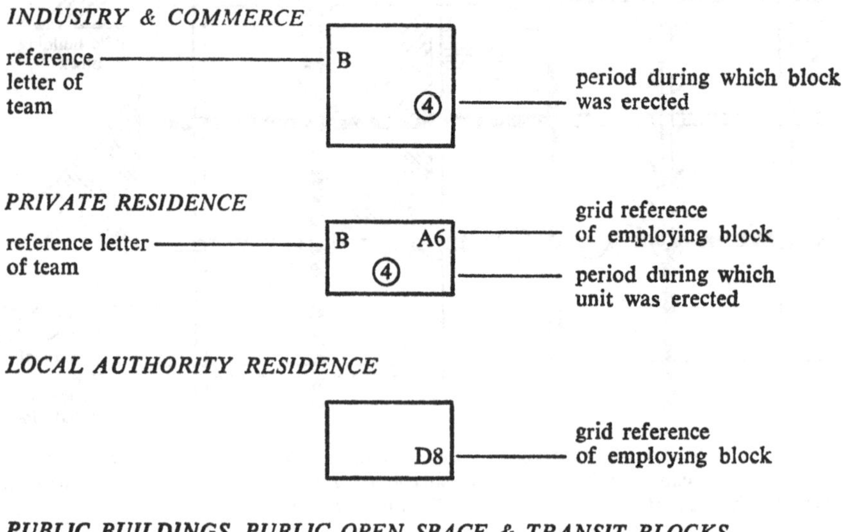

PUBLIC BUILDINGS, PUBLIC OPEN SPACE & TRANSIT BLOCKS

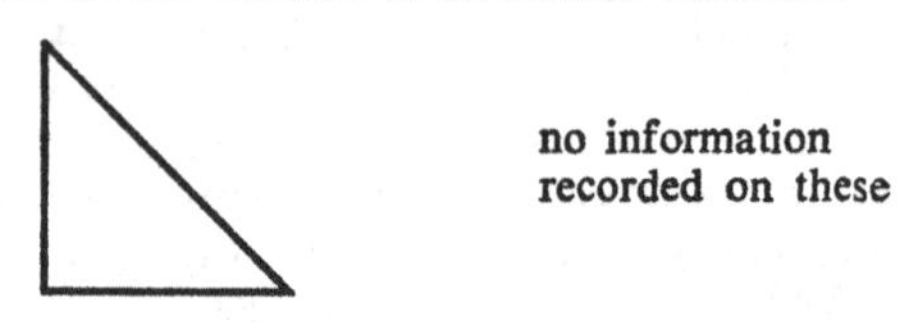

FORM 8. COST OF BUILDING BLOCKS (STAGE 12)

Blocks built in any one period (or any one type only)	Industry I	Commerce C	Residence R
1	5,000	2,500	1,000
2	4,500	2,300	1,000
3	4,000	2,000	900
4	4.000	1,800	900
5 or more	4,000	1.800	800

FORM 9. INCOME FROM ACTIVITY BLOCKS (STAGE 13)

	Values for D									
Net income formula	1	2	3	4	5	6	7	8	9	10
INDUSTRY D measured to transit block 2.000—(400 × D)	1,600	1,200	800	400	—	—	—	—	—	—
COMMERCE D measured to transit block 900—(100 × D)	800	700	600	500	400	300	200	100	—	—
RESIDENTIAL D measured to employing block 250—(10 × D)	240	230	220	210	200	190	180	170	160	150

FORM 10. PERIOD RECORD SHEET

Team .... Period .... Time .... Date	Debit	Credit	Balance
1. Enter cash balance at start of period			+
2. Enter interest on cash balance (=10% of item 1 of last period)		+	
3. Consider development plan proposals (as appropriate)			
4. Negotiate loans as required. Cash loaned in multiples of 4,000. Loan repayments in multiples of 1,000 over next 5 periods	–	+	
5. Construct new communication links as required. A majority vote: each link adds 30 to the total rates			
6. Construct new public buildings (health, schools, etc.). A majority vote: each new building adds 30 to the total rates			
7. Construct new public open spaces as required. A majority vote: each new block adds 30 to the total rates			
8. Construct new transit blocks as required. A majority vote: each new block adds 300 to the total rates			
9. Calculate the cost of contingencies (gales, floods, fires, etc.). On periods 4, 8, 12, etc. 2 dice thrown to determine loss	–		
10. Commuter deductions as applicable No. of residential blocks without commercial services ____ No. of residential blocks without public open space ____ No. of residential blocks without public buildings ____ Total shortfall in facilities × 50 gives deduction (if any facilities are not on board, then double shortfall for that one)	–		
11. Buy or sell land — Enter details below, one column per plot		+	
12. Construct buildings			
Plot			
Cost if bought now	–		
Building type			
Building cost	–		
Price of land sold			
13. Net income from new buildings built above I 2,000−D to T×400 + C 900−D to T×100 + R 250−D to W× 10 +			
14. Continuing net income from existing buildings (i.e. (15) of last period) − income from buildings demolished= +			
15. New total net income from all buildings .... +		+	
16. Calculate rates assessment Rate per plot ______ × number of plots you occupy ______	–		
17. Balance at end of period			

All figures are handled in £1,000 units, e.g. 30 here means £30,000 approx.

FORM 11. RATES RECORD SHEET

Period / Time		No. of communication links	No. of public buildings	No. of POS blocks	No. of transit blocks × 10	A+B+C+D	E × 30	No. of plots occupied by private teams	Rate per plot
		A	B	C	D	E	F	G	F/G
Brought Forward									
	New								
	NEW TOTAL		+	+	+	=	×30=		
	New								
	NEW TOTAL		+	+	+	=	×30=		
	New								
	NEW TOTAL		+	+	+	=	×30=		
	New								
	NEW TOTAL					=	×30=		
	New								
	NEW TOTAL		+	+	+	=	×30=		
	New								
	NEW TOTAL		+	+	+	=	×30=		

FORM 12. BASIC ROLE RELATIONSHIPS

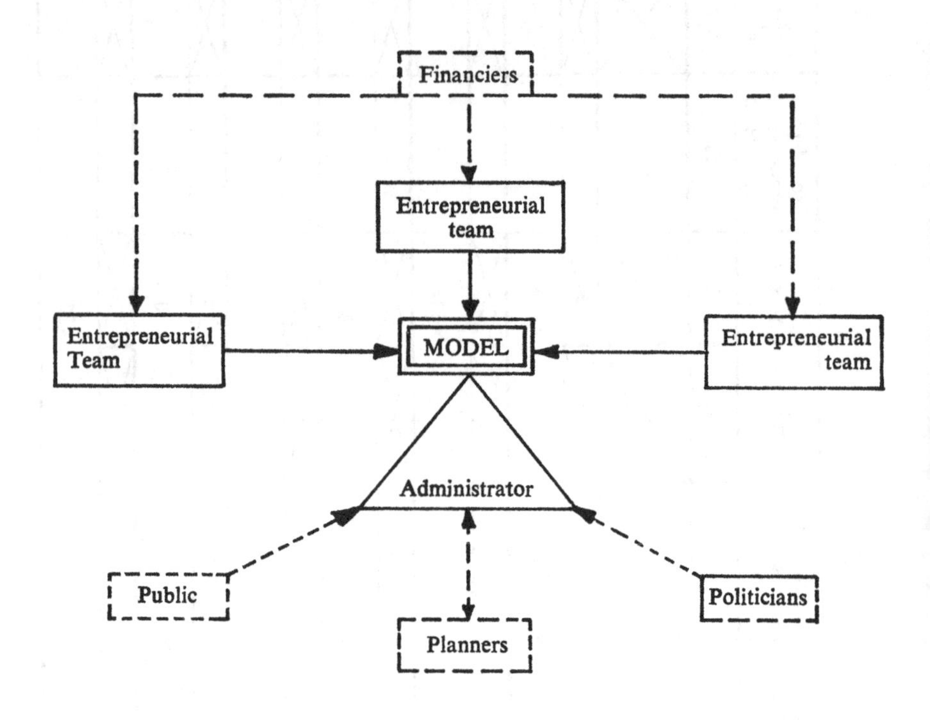

ROLE	MINIMUM NUMBER OF TEAMS	MINIMUM SIZE OF TEAMS
Entrepreneurs	3	1
Administrators	1	1
Planners	1	1
Public	1	1
Politicians	1	1
Financiers	1	3

Appendix 2 A directory of selected urban development simulations

Title	Description	Operational requirements	Level	Designers/Suppliers
APEX	Air pollution training exercise which is an outgrowth of the METRO simulation model	17+players (min.) 9 hours (min.) Computer essential	Undergraduate through to professional	Dr R. D. Duke, Environmental Simulation Lab., University of Michigan, Ann Arbor, Michigan 48104, U.S.A.
BUILD (a community development game)	Computer-based decision-making exercise involving extensive role playing which was designed to help develop effective participation in the planning of new communities	17+players (min.) 17+ hours Computer essential	General public through to professional and technical	Prof. A. J. Pennington, Drexel Institute of Technology, Philadelphia, Pennsylvania 19104, U.S.A.
CITY I	Computerized metropolitan simulation which includes social relationship and political role playing in relation to a central city area as well as 3 suburban areas	17+ players 9–16 hours Computer essential	Senior school and above	Dr Peter House, Envirometrics, 1100 17th St. N.W., Washington D.C. 20036, U.S.A.
CLUG (COMMUNITY LAND USE GAME)	Analogue of the interactions and changes involved in urban land use planning at a high order of abstraction. It is a 'board' type of game, with specific and fairly rigid rules, which might be compared to a combination of chess and monopoly	9 players (min.) 4 hours (min.) Computer optional	Senior school and above	Dr A. G. Feldt, Dept. City and Regional Planning, Cornell University, Ithaca, N.Y. 14850, U.S.A. Kit available from: System Gaming Associates, Triphammer Road, Ithaca, N.Y. 14850, U.S.A.

Title	Description	Operational requirements	Level	Designers/Suppliers
COMEXOPOLIS	Early adaptation of the METROPOLIS model incorporating a smog factor (see APEX above)	10 players (min.) 2–4 hours (min.) Computer essential	Undergraduate and above	Air Pollution Computer Research Unit, University of Southern California, Los Angeles, U.S.A.
DE KALB	Urban mayoral election simulation	17+ players 2–4 hours	Undergraduate and above	Prof. M. Whithead, Political Science Department, Temple University, Philadelphia, Penn. 19122, U.S.A.
ECOGAME	Computerized video display game demonstrating resource allocation problems in a simple socio-economic system	4 teams (min.) ½ hour (min.) Computer essential	General educated public	G. L. Mallen, Systems Research Ltd, 20 Hill Rise, Richmond, Surrey, U.K.
ECONOMIC DEVELOPMENT SIMULATION MODEL	Decision-making exercise centred on a mathematical model of a hypothetical less developed economy	2 players (min.) 5 hours (min.) Computer essential	Postgraduate and professional	C. L. G. Bell, Institute of Development Studies, University of Sussex, Falmer, Nr. Brighton, Sussex, U.K.
EDUCATION SYSTEM PLANNING GAME	'Human-player' simulation of the planning, programming and budgeting process in relation to education	17+ players 2–4 hours (min.)	Undergraduate and above	Abt Associates Inc., 14 Concord Lane, Cambridge, Mass., U.S.A.
EXERCISE QUINTAIN	Management exercise covering some of the economic aspects of a typical building and civil engineering firm. The model is Laing's version of Operation Taurus	5+ players 5–8 hours	Undergraduate and above	John Laing and Sons Ltd, London, N.W.7, U.K.
GAME SIMULATE OF THE ARCHITECTURAL CONTROL PROCESS	Programmed decision-making exercise relating to aesthetic and related planning controls used in the regulation of development	10+ players 4 hours	Undergraduate and above	Dr Sidney Cohn, Dept. City and Regional Planning, University of North Carolina, Chapel Hill, N.C. 27514, U.S.A.

Title	Description	Operational requirements	Level	Designers/Suppliers
MANAGEMENT PLANNING GAME (GSPIA)	Set of interlocking decision-making exercises structured around a 'nest' of urban models	17+ players 17+ hours Computer essential	Graduate and research	Prof. Clark Rogers, Dept. of Urban Affairs, University of Pittsburgh, Pittsburgh, Penn. 15213, U.S.A.
ILAG (InLoGov LOCAL AUTHORITY GAME)	Local government management exercise with various elaborations developed to be applicable to particular situations and the Institutes expanding 'nest' of models.	10 players 6 hours (min.)	Undergraduate and above	R. H. R. Armstrong, Institute of Local Government Studies, University of Birmingham, Box 363, Birmingham B15 2TT, U.K.
IMPACT	Simulation designed to illustrate how individual and collective actions affect an imaginary community	25–30 players 5–8 hours	Senior school and above	Dr R. G. Klietsch, 3M Co. (Learning Production), 3M Center, St Paul, Minnesota 55101, U.S.A.
INSIGHT	Serialized simulation involving four distinct but related units, each focusing on problem situations in a multiracial community	17+ players 9–16 hours	Senior school and above	Instructional Simulation Inc., Box 212, Newport, Minnesota 55055, U.S.A.
LUGS (LAND USE GAMING-SIMULATION)	Anglicized derivative of the basic CLUG model involving a simple simulation of the urban-development process in terms of selected and well-defined characteristics (see Appendix 1)	4+ players 4 hours (min.)	Senior school and above	Dr J. L. Taylor, Depart. Town and Regional Planning, University of Sheffield, Sheffield S10 2TN, U.K.
LAND USE–TRANSPORTATION SIMULATION	'Human player' simulation designed to provide opportunities to experience, analyse and compare the long-term effect of various planning policies in a fictitious town with a population of 50,000	13–35 players 8 hours (min.) Packaged materials	Undergraduate and above	P.T.R.C. Co. Ltd, Grosvenor Gardens, London, S.W.1, U.K.

Title	Description	Operational requirements	Level	Designers/Suppliers
LOW BIDDER	Simple packaged business game which simulates construction bidding procedures involved in the competitive tendering process	2+ players 1 hour	Middle school and above	Enterlek Inc., 42 Pleasant Street, Newburyport, Mass. 01950, U.S.A.
MANCHESTER	Simulation of the impact on the agricultural population of some of the major historical and social issues surrounding the advent of the industrial revolution in England	8–40 players 1–2 hours	Senior school and above	Abt Associates Inc., 14 Concord Lane, Cambridge, Mass., U.S.A.
METRO	Large computerized urban simulation designed as an outgrowth from the METROPOLIS model as a training and research operation	17+ players 9 hours (min.) Computer essential	Postgraduate	Dr R. D. Duke, Environmental Simulation Lab., University of Michigan, Ann Arbor, Michigan 48104, U.S.A.
METROPOLIS	Community development simulation designed to familiarize the players with some of the more significant decision-making roles affecting urban growth	9–16 players 4 hours (min.) Computer assistance an advantage	Undergraduate and above	As above
NEIGHBORHOOD	Simulation of aspects of urban development at the local level involving problems of physical planning and social organization	4–12 players 1–2 hours	Middle school and above	Wellesley School Curriculum Center, 12 Seaward Road, Wellesley Hills, Mass. 02181, U.S.A.
N.E. CORRIDOR TRANSPORTATION GAME	Computer-assisted role-playing simulation using information from the 'Corridor' between Washington D.C. and Boston	17+ players 17+ hours Computer and related electronic assistance required	Professional and research	Abt Associates Inc., 14 Concord Lane, Cambridge, Mass., U.S.A.

Title	Description	Operational requirements	Level	Designers/Suppliers
NEW TOWN	Simple and engaging packaged land use simulation game design to introduce to a wide audience some of the forces shaping urban development	4+ players 2–4 hours	Senior school and above	'New Town', 1108 North Troga Street, Ithaca, N.Y. 14850, U.S.A.
OPERATION TAURUS	Management exercise based on the economic aspects of a typical building and civil engineering firm	9–16 players 5–15 hours	Undergraduate and above	P. C. Webb and G. E. Wheeler, Dept. of Management Studies, Hendon Technical College, Hendon, Middlesex, U.K.
OPERATION SUBURBIA	Simple exercise involving land dealing and real-estate development designed to demonstrate how groups often fail to plan or organize themselves	9–35 players 1–2 hours	Senior school and above	Dr A. A. Zoll. Management Education Assoc., 2003 Thirty-third Avenue South, Seattle, Washington 98144, U.S.A.
ORL-PLANUNGSSPIEL	Swiss derivative of the English 'Land Use Gaming-Simulation' (LUGS) model which in addition owes much to the Community Land Use Game in terms of format and procedures	10+ players 2–4 hours Computer optional	Undergraduate and above	M. Geiger, ORL Institut, Erdgenossischen Technissche Hochschule, Zurich, Switzerland
POGE (PLANNING OPERATION GAMING EXPERIMENT)	Demonstration of gaming techniques applied to the interaction between property developer and planners	3+ players 2–4 hours	Undergraduate and above	Prof. F. Hendricks, School of Architecture, California State Polytechnic College, California 93401, U.S.A.
PLANS	Influence allocation game simulating the relationships between military, civil rights, national, international, business and labour interests in American society	12–30 players 3–8 hours	Senior school and above	Western Behavioral Sciences Institute, Project Simile, P.O. Box 1023 1150 Silverado, La Jolla, Cal. 92037, U.S.A.

Title	Description	Operational requirements	Level	Designers/Suppliers
POLICY NEGOTIATION GAME	Resource allocation exercise capable of simulating numerous situations and having the added advantage of being open to continuous re-norming procedures	5 players (min.) 2 hours Inexpensive operational unit	Senior school, general public and above	Dr F. L. Goodman, School of Education, University of Michigan, Ann Arbor, Michigan 48104, U.S.A.
POLIS: URBAN EXERCISE	Role-playing simulation, involving the major political elements in the city of San Diego, California, which 'interlocks' with other sectoral, political models	30–50 players 9–30 hours Computer assistance optional	Undergraduate and research	Prof. R. C. Noel, Dept. of Political Science, University of California, California 93106, U.S.A.
POLLUTION	Simulation covering some of the major social, political, and economic problems involved in attempts to control pollution	12–24 players 2–4 hours	Middle school and above	Wellesley School Curriculum Center, 12 Seaward Road, Wellesley Hills, Mass. 02181, U.S.A.
PRIORITY EVALUATOR GAME	Development of a research game designed by the Institute of Research in Social Science, University of N. Carolina, and used to gauge community environmental preferences	1 player 30–40 minutes	General public	G. Hoinville, Social and Community Planning Research, 41 Doughty Street, London, W.C.1, U.K.
REGION	Computerized simulation which provides a regional view of urban growth with a strong economic base theory orientation	17+ players 5–8 hours Computer assistance essential	Undergraduate and above	Dr P. House, Envirometrics, 1100 17th St. N.W., Washington D.C. 20036, U.S.A.
REGIONAL PLANNING GAME	Regional decision-making game which portrays the interaction between certain key Scottish communities (Aberdeen, Inverurie, Huntly, Keith and Elgin) as well as between major 'influencing' agents such as government and industry	9–16 players 17+ hours	Undergraduate and above	A. Henderson, Town and Regional Planning School, Duncan of Jordanstone College of Art, Dundee, Scotland

Title	Description	Operational requirements	Level	Designers/Suppliers
ROUTE LOCATION GAME	Role-playing simulation design to uncover, as well as weight, various community values and aims to provide a greater appreciation of problems and decisions faced in urban planning as shown when considering the delineation of an alignment for a public highway	17+ players 5–8 hours	General public and university	Creighton, Hamburg Inc., 231 Delaware Avenue, Delmar, N.Y. 12054, U.S.A.
MICRO-MODEL (SFCRP)	Manual game designed originally to train professional staff to operate the computerized model of the San Francisco Housing Market developed by Arthur D. Little	5–8 players 5 hours	Professional planning	Prof. F. Hendricks, School of Architecture, California State Polytechnic College, California 93401, U.S.A.
SECTION	Simulation designed to provide students with an understanding of conflicts of interest among the sections of various political territories as they are expressed in the political process	30+ players 5–6 hours	Senior school and above	Abt Associates Inc., 14 Concord Lane, Cambridge, Mass., U.S.A.
SIMULAND	Ambitious simulation representing the political, social, and economic processes in a developing country that is patterned after a Latin American model	17+ players 9–16 hours	Undergraduate and above	Prof. A. M. Scott *et al.*, Dept. of Politics, University of N. Carolina, Chapel Hill, N.C. 27514, U.S.A.
SIMSOC (SIMULATED SOCIETY)	One of the better documented urban community simulations fully described with background reading in Gamson's 1969 Collier Macmillan Publications	17+ players 9–16 hours	Undergraduate and above	Dr W. A. Gamson, Dept. of Sociology, University of Michigan, Ann Arbor, Michigan 48104, U.S.A.

Title	Description	Operational requirements	Level	Designers/Suppliers
SITTE	Influence allocation game simulating the impact of five interest groups on the changing quality of life in the mythical city of SITTE	17+ players 2–4 hours	Middle school and above	Western Behavioral Sciences Institute, Project Simile. P.O. Box 1023, 1150 Silverado, La Jolla, Cal. 92037, U.S.A.
STARPOWER	Societal advancement game which simulates community mobility and power structures and is available as a packaged kit	18–35 players 2–4 hours	Middle school and above	As above
SUMERIAN GAME	Simulation requiring the student to take the role of town ruler about 3500 B.C., faced with a variety of elementary community development problems	1 player 5–8 hours Computer and multi-media facilities required	Middle school and above	BOCES, Center for Educational Services and Research, 845 Fox Meadow Road, Yorktown Heights, N.Y. 10598, U.S.A.
SUNSHINE	Simulation in which each participant is 'born' into the community of a town called SUNSHINE: different racial and community identities are assumed and the game revolves around various community development problems and their impact on roles and relationships	10–35 players 9–16 hours	Middle school and above	'Interact', P.O. Box 262, Lakeside, California 92020, U.S.A.
SYSTEMS ANALYSIS MODEL OF URBANIZATION AND CHANGE	Very flexible simulation of alternative planning decision chains developed in an educational context	17+ players 2–4 hours Computer assistance optional	Graduate and research	Prof. C. Steinitz *et al.*, Harvard Graduate School of Design, Cambridge, Mass. 02138, U.S.A.

Title	Description	Operational requirements	Level	Designers/Suppliers
TELECITY	Computer-assisted model owing much to Envirometrics' other gaming simulations and simulating decisions affecting the economic, social and governmental conditions in a metropolitan area	20–100 players 5–8 hours Computer essential	Senior school and above	Dr P. House, Envirometrics, 1100 17th St. N.W., Washington D.C. 20036, U.S.A.
TORONTO GAME	Simulation centred on Don Mills. Ontario, which portrays some of the major role interactions in urban settlements in the twentieth century	4–17+ players 2–4 hours	Undergraduate and above	Prof. S. N. Benjamin, Department of Architecture, University of Toronto, Toronto, Canada
TRACTS	Packaged socio-political simulation illustrating some of the dynamics of land value, ownership and land use politics in a simulated core city area	17+ players 2–4 hours	Senior school and above	Instructional Simulations Inc., Box 212, Newport, Minnesota 55055, U.S.A.
TRADE-OFF	Budget allocation simulation viewed by its designers as a data collection device with educational potential in the extra-mural area	Unlimited no. of players 2+ hours	General public	J. Berger and Dr L. K. Walford, La Clede Town Co., St. Louis, Missouri 63103. U.S.A.
URBAN PLANNING SIMULATION	Framework for culling user information from individuals and groups who find it difficult to articulate personal as well as community needs and desires	2–4 players 2–4 hours	Undergraduate and above	Prof. N. B. Mitchell, 149 Putnam Avenue, Cambridge, Mass. 02139, U.S.A.
VIRGIN ISLAND GAME	Simulation sponsored by the College of the Virgin Islands to clarify some of the internal communication patterns amongst government and the community in a developing nation	30+ players 2–4 hours (min.)	General public	Abt Associates Inc., 14 Concord Lane, Cambridge, Mass., U.S.A.

Title	Description	Operational requirements	Level	Designers/Suppliers
WILDLIFE	Series of games simulating various aspects of population dynamics through an examination of the concept of community in its elementary ecological sense	2–4 players 2 hours (min.) Computer assistance optional	Senior school and above through to research	Dr R. L. Meier, School of Natural Resources, University of Michigan, Ann Arbor, Michigan 48104, U.S.A.
WOODBURY	Political campaign simulation about to be published by Little, Brown and Co. (U.S.A.)	17+ players 2–4 hours	Undergraduate and above	Prof. M. Whithead, Political Science Dept., Temple University, Philadelphia, Penn. 19122, U.S.A.

Bibliography

Abt, C. C. (1964). ' War Gaming ', *International Science and Technology,* no. 32, August, pp. 29–37.

Abt Associates Inc. (1965*a*). ' Report of a Survey of the State of the Art: Social, Political, and Economic Models and Simulations ', for the National Commission on Techology, Automation, and Economic Progress, Washington D.C., Cambridge, Mass.: Abt Associates Inc. (mimeo).

Abt, C. C. (1965*b*). ' An Educational System Planning Game ', Cambridge, Mass.: Abt Associates Inc. (mimeo).

Abt, C. C. (1966). ' Games for Learning ', Cambridge, Mass.: The Social Studies Curriculum Project, Educational Services Incorporated, Occasional Paper no. 7.

Abt Associates Inc. (1967*a*). ' North East Corridor Transportation Game: Game Administrator Handbook ', Cambridge, Mass.: Abt Associates Inc. (mimeo).

Abt Associates Inc. (1967*b*). ' Simpolis, A Metro Game ', Cambridge, Mass.: Abt Associates Inc. (mimeo).

Abt Associates Inc. (1968*a*). ' Final Report on the Virgin Islands Game ', Cambridge, Mass.: Abt Associates Inc.

Abt, C. C. (1968*b*). ' The Impact of Technological Change on World Politics ', *The Futurist,* vol. 11, no. 2, April, pp. 21–5.

Alexander, C. (1965). ' The Co-ordination of Urban Rule Systems ', in *1965 Internationale Regio Planertagung,* Regio: Basel, pp. 168–76.

American Management Association (1961). *Simulation and Gaming: A Symposium,* New York: A.M.A. Inc., General Management Division.

Anderson, L. F., Herman, M. G., Robinson, J. A. and Snyder, R. C. (1964). *A Comparison of Simulation, Case Studies, and Problem Papers in Teaching Decision Making,* Co-operative Research Project, no. 1568, Northwestern University, Evanston, Illinois.

Anderson, R. A. (undated). ' Instructional Unit on the Use of CLUG ', Developed by the Urban-Regional Research Institute, Michigan State University, under the auspices of the Ford Foundation (Director of U.R.R.I.—Richard D. Duke).

Andlinger, G. R. (1958). ' Business Games—Play One ', *Harvard Business Review,* XXXVI, no. 2 (1958), pp. 115–25.

Angell, N. W. (1928). *The Money Game: A New Instrument of Economic Education,* London: J. M. Dent.

Anon (1968). ' Urban Planners Play a Game Called City I ', *Business Work*, November 16, pp. 66–70.

Armstrong, R. H. R. (1968). ' The Inlogov Local Authority Game: Background Paper and Commentary ', Institute of Local Government Studies, University of Birmingham, June (mimeo).

Armstrong, R. H. R. (1970*a*). ' The Use of Operational Gaming in the field of Local Government Studies ', in Armstrong, R. H. R. and Taylor, J. L. (eds.), *Instructional Simulation Systems in Higher Education*, pp. 67–111.

Armstrong, R. H. R. (1970*b*). ' Operational Gaming and P.P.B.S.', in *P.P.B.S. and Decision Making in Local Government*, P.T.R.C. Proceedings, vol. 2, pp. 15–16, London: P.T.R.C.

Armstrong, R. H. R. and Hobson, M. (1969*a*). ' Games and Urban Planning ', *Surveyor*, 31 October, pp. 32–4.

Armstrong, R. H. R. and Hobson, M. (1969*b*). ' Models for Life ', *Education*, vol. 134 (10), September, p. 1087.

Armstrong, R. H. R. and Hobson, M. (1969*c*). ' " Planning Games " are more than just fun ', *Municipal and Public Services Journal*, vol. 2089, 7 November.

Armstrong, R. H. R. and Hobson, M. (1969*d*). ' ALEA—Manual and Rules for the Blackburn-Leyland Chorley Model A ', Birmingham: Institute of Local Government Studies (mimeo).

Armstrong, R. H. R. and Hobson, M. (1970). ' Participation Through Games ', *Town and Country Planning*, vol. 38, no. 2, February, pp. 125–6.

Armstrong, R. H. R. and Taylor, J. L. (eds.) (1970). *Instructional Simulation Systems in Higher Education*, Cambridge Institute of Education: Cambridge Monographs on Education, no. 2.

Armstrong, R. H. R. and Taylor, J. L. (eds.) (1971). *Feedback on Instructional Simulation Systems*, Cambridge Institute of Education: Cambridge Monographs on Education, no. 5.

Association of American Geographers (1967). *Growth of Cities Unit : High School Geography Project*, Boulder, Colorado: Association of American Geographers.

Association of Teachers of Management (1965). *Business Exercises*, Occasional Paper no. 1, Oxford: Basil Blackwell.

Attiyeh, R. and Brainard, W. (1968). ' A Simulation Policy Game for Teaching Macroeconomics ', *American Economic Review*, vol. LVIII, no. 2, May, pp. 458–68.

Auld, H. D. and Crawford, J. D. (1963). ' A Note on Management Games ', *Occupational Psychology*, vol. 37, no. 2, April, pp. 130–7.

Banks, M. H., Groom, A. J. R. and Oppenheim, A. N. (1968). ' Gaming and Simulation in International Relations ', *Political Studies*, vol. XVI, no. 1, February 1968, pp. 1–17.

Banks, M. H., Groom, A. J. R. and Oppenheim, A. N. (1970). 'Gaming, Simulation and the Study of International Relations in British Universities', in Armstrong, R. H. R. and Taylor J. L. (eds.), *Instructional Simulation Systems in Higher Education*, pp. 28–44.

Bayliss, D. (1968). 'Some Changing Characteristics of Research in Environmental Studies', London: Centre for Environmental Studies, Working Paper CES-WP-U, July.

Beale, E. M. L. (1961). 'The Role of Gaming in Military Operational Research', D.O.R. Memorandum, no. 190 (ACSIL/ADM/61/28).

Beard, R. M. (1967). 'Using Tests to aid Learning', in *Teaching for Efficient Learning*, A Report of the proceedings at the second conference organised by the University Teaching Methods Research Unit held at the Institute of Education, London, 6 January.

Beard, R. M. (1970). 'Some Perspectives on Research in Higher Education', in Armstrong, R. H. R. and Taylor, J. L. (eds.), *Instructional Simulation Systems in Higher Education*, pp. 1–8.

Beard, R. M. (1970). *Teaching and Learning in Higher Education*, Harmondsworth, Middlesex: Penguin Education.

Bell, C. (1970). 'A Development Game: Decision Making in a Simulated Less Developed Economy', in Armstrong, R. H. R. and Taylor, J. L. (eds.), *Instructional Simulation Systems in Higher Education*, pp. 127–34.

Bellman, R., Clark, C. E., Malcolm, D. G., Craft, C. J. and Ricciardi, F. M. (1957). 'On the Construction of a Multi-Stage Multi-Person Business Game', *Operations Research*, August.

Benjamin, S. (1968). 'Operational Gaming in Architecture', *Architecture Canada*, vol. 45, no. 2, February.

Berkeley, E. P. (1968). 'The New Gamesmanship—a report on the new urban games', *Architectural Forum*, vol. 129, no. 5, December, pp. 58–63.

Bernholtz, A. (1968). 'Some Thoughts on Computer Role Playing and Design', *Ekistics*, vol. 26, no. 157, December, pp. 522–4.

Bernholtz, A. (1968). 'Some Thoughts on Computer Role Playing and Design', *Connection*, vol. 5, nos. 2/3, winter/spring, pp. 88–91.

Blaxall, J. (1965). 'Manchester', Cambridge, Mass.: Abt Associates (mimeo).

Blivice, S. (1970). 'Simulation Review: The Community Land Use Game', *Simulation and Games*, vol. 1, no. 2, June, pp. 221–4.

Bloom, P. (1969). 'Gaming Simulation in Planning Education', Master of Regional and Community Planning Thesis, Manhattan, Kansas: Kansas State University.

Bloomfield, L. P. (1960). 'Political Gaming', *U.S. Naval Institute Proceedings*, September, pp. 57–64.

Bloomfield, L. P. and Padelford, N. J. (1959). 'Three Experiments in Political Gaming', *The American Political Science Review*, LIII, pp. 1105–15.

Bloomfield, L. P. and Whaley, B. (1965). 'The Political-Military Exercise: A Progress Report', *Orbis*, Foreign Policy Research Institute, University of Pennsylvania, pp. 854–70.

Boguslaw, R., Davis, R. H. and Glick, E. B. (1966). *PLANS: Participants Manual.* La Jolla, California: Western Behavioral Sciences Institute.

Boocock Sarane, S. and Schild, E. O. (eds.) (1968). *Simulation Games in Learning*, Beverly Hills, California: Sage Publications Inc.

Borger, R. and Seaborne, A. E. M. (1966). *The Psychology of Learning*, Harmondsworth, Middlesex: Penguin Books.

Bourgeois, D. A. (1969). 'Planning for the Model City in St. Louis', in Campbell, R. F., Marx, L. A. and Nystrand, R. O. (eds.), *Education and Urban Renaissance*, New York: John Wiley and Sons Inc.

Braybrooke, D. and Lindblom, C. E. (1963). *A Strategy of Decision*, Glencoe, Illinois: Free Press.

British Institute of Management (1962). *Management Training Techniques*, London: B.I.M. publication.

Brody, R. A. (1963). 'Varieties of Simulations in International Relations Research', in *Simulation in International Relations: Developments for Research and Teaching*, Englewood Cliffs, N.J.: Prentice Hall, pp. 190–223.

Bruce, R. L. (1967). 'A Layman's Introduction to CLUG I', Ithaca, New York: School of Education, Cornell University (mimeo).

Bruner, J. S. (1961). *The Process of Education*, Cambridge, Mass.: Harvard University Press.

Bruner, J. S. (1966). *Toward a Theory of Instruction*, Cambridge, Mass.: Harvard University Press.

Burgess, P. M. (1966). 'Social Studies and Social Skills; The Education of Political Man', paper prepared for delivery at the 46th Annual Meeting of the National Council for Social Studies in Joint Session with the American Political Science Association, Cleveland, Ohio, November 26.

Caillois, R. (1961). *Man, Play and Games*, New York: Free Press.

Carter, K. R. (1969). 'Survey of Student Reaction to the Cornell Land Use Game', Coventry: Lanchester College of Technology, Department of Town Planning (unpublished paper).

Carter, K. R. and Taylor, J. L. (1968). 'Gaming-Simulation', Report of the Proceedings of the Town and Country Planning Summer School held at the University of Manchester, September, pp. 62–3.

Chadwick, G. F. (1966). 'A Systems View of Planning', *Town Planning Institute Journal*, vol. 52, no. 5 pp. 184–6.

Cherryholmes, C. H. (1966). 'Some Current Research on Effectiveness of Educational Simulations: Implications for Alternative Strategies', *American Behavioral Scientist*, vol. 10, no. 2, October.

Chorley, R. J. and Haggett, P. (eds.) (1967). *Models in Geography*, London: Methuen.

Cohen, K. J. and Rhenman, E. (1961). 'The Role of Management Games in Education and Research', *Management Science*, VII, January, pp. 131–66.

Cohen, K. J., Dill, W. R., Kuehn, A. and Winters, P. (1964). *The Carnegie Tech Management Game: An Experiment in Business Education*, Homewood, Illinois: Richard D. Irwin, Inc.

Cohn, S. (1968). 'Simulating the Architectural Control Process', Chapel Hill, North Carolina: Department of City and Regional Planning, University of North Carolina (mimeo).

Cole, J. P. and Beynon, N. J. (1969). *New Ways in Geography*, Oxford: Basil Blackwell.

Cole, K. D. O. (1961). 'Role-Playing', in *Management Training Techniques*, Proceedings of a B.I.M. Conference, 8 June 1961, London: British Institute of Management.

Coleman, J. S. (1961). *The Adolescent Society*, New York: Free Press.

Coleman, J. S. (1965). *Adolescents and the Schools*, New York: Basic Books.

Coleman, J. S. (1968*a*). 'Games as Vehicles for Social Theory', Report no. 21, Johns Hopkins University, Centre for the Study of Social Organization of Schools, Baltimore, Maryland: Johns Hopkins University, May (mimeo).

Coleman, J. S. (1968*b*). 'Social Processes and Social Simulation Games', in Boocock, S. S. and Schild, E. O. (eds.), *Simulation Games in Learning*, Beverly Hills, California: Sage Publications Inc., pp. 29–52.

Coleman, J. S., Boocock, S. S. and Schild, E. O. (eds.) (1966). 'Simulation Games and Learning Behavior', *American Behavioral Scientist*, vol. 10, nos. 2 and 3 (October and November).

Cooke, J. E. (1970). 'Experience Training: Attitude Changing by Means of Undergoing an Experience', in Armstrong, R. H. R. and Taylor, J. L. (eds.), *Instructional Simulation Systems in Higher Education.*

Cooper, E. (1967). 'Games with a Purpose', *Works Management*, vol. 20, no. 5, May.

Crawford, M. P. (1965). 'Dimensions of Simulation', Presidential Address, Division of Military Psychology (Div. 19), 73rd Annual Convention of the American Psychological Association, 6 September 1965, Washington: Human Resources Research Office, George Washington University.

Crecine, J. P. (1964). 'Time-Oriented Metropolitan Model', C.R.P. Technical Bulletin, no. 5, Pittsburgh City Planning Department, January.

Crecine, J. P. (1968). 'A Dynamic Model of Urban Structure', Paper P-3803, Santa Monica, California: Rand Corp., March.

Creighton, Hamburg, Inc. (undated). ' Route Location Game ', Delmar, New York (mimeo).

Crow, W. J. and Noel, R. C. (1965). ' The Valid Use of Simulation Results ', La Jolla, California: Western Behavioral Sciences Institute (mimeo).

Cushen, W. E. (1966). ' The Operational Research Group in the Department of Commerce: The North East Corridor Transportation Study ', Proceedings of the Symposium on the Contribution of Operational Research to Urban and Regional Planning, Rome, 5–7 December 1966.

Dale, A. G. and Klasson, C. R. (1964). *Business Gaming : A Survey of American Collegiate Schools of Business*, Bureau of Business Research, The University of Texas, Austin.

Davidoff, P. (1965). ' Advocacy and Pluralism in Planning ', *American Institute of Planners Journal*, vol. 31, no. 4, pp. 331–8.

Davison, W. P. (1961). ' A Public Opinion Game ', *Public Opinion Quarterly*, 25, pp. 210–20.

Dawson, R. E. (1962). ' Simulation in the Social Sciences ', in Guetzkow, H. ed.), *Simulation in Social Science*, Englewood Cliffs, N.J.: Prentice Hall Inc.

Dearden, R. F. (1968). *The Philosophy of Primary Education : An Introduction*, London: Routledge and Kegan Paul.

Defence Operational Analysis Establishment (undated). ' Glossary of Terms used in Gaming and Simulation ', West Byfleet, Surrey: D.O.A.E.

Design Methods Group (1969). ' Questionnaire ', *Design Methods Group Newsletter*, vol. 3, no. 12, December, pp. 3–10.

De Sola Pool (1964). ' Simulating Social Systems ', *International Science and Technology*, March, pp. 62–70.

Dewey, J. (1922). *Human Nature and Conduct*, New York: Henry Holt.

Dill, W. R., Jackson, J. R. and Sweeney, J. W. (eds.) (1961). *Proceedings of the Conference on Business Games as Teaching Devices*, The Ford Foundation and School of Business Administration, Tulane University, New Orleans, Louisiana, 26–8 April.

Dill, W. R. and Doppelt, N. (1963). ' The Acquisition of Experience in a Complex Management Game ', *Management Science*, vol. 10, no. 1, October, pp. 30–46.

Domitriou, B. (1971). ' Video Tape Recording in Gaming Simulation ', in Armstrong, R. H. R. and Taylor, J. L. (eds.), *Feedback on Instructional Simulation Systems*, Cambridge Institute of Education: Cambridge Monographs on Education no. 5.

Dotson, Z. B. and Sawicki, D. S. (1968). ' Systems Gaming Associates Announcement ', Ithaca, New York (mimeo).

Duke, R. D. (1964). *Gaming Simulation in Urban Research*, Institute for Community Development, Michigan State University, East Lansing, Michigan.

Duke, R. D. (1965). 'Simulation for Planning: Gaming Urban Systems', *Planning 1965*, selected papers from the 1965 Joint Planning Conference of the American Society of Planning Officials and the Community Planning Association of Canada, Toronto, Canada, 25--9 April.

Duke, R. D. (1967). 'Planning Research at Michigan State University: an interview with John L. Taylor', *Journal of the Town Planning Institute*, vol. 53, no. 6, pp. 239–41.

Duke, R. D. and Schmidt, A. H. (1965). 'Operational Gaming and Simulation in Urban Research: An Annotated Bibliography', Institute for Community Development, Continuing Education Service, Michigan State University, Bibliography no. 14, January.

Duke, R. D. *et al.* (1966). 'M.E.T.R.O. Project Technical Report no. 5: Report on Phase 1', Tri-County Regional Planning Commission, Lansing, Michigan, January.

Duke, R. D. and Burkhalter, B. R. (1966). 'The Application of Gaming to Urban Problems', Institute for Community Development, Continuing Education Service, Michigan State University, Technical Bulletin B-52, January.

Dyckman, J. W. (1966). 'Social Planning, Social Planners and Planned Societies', *Journal of the American Institute of Planners*, vol. 32, no. 2, March, pp. 66–76.

Edwards, R. F. and Francis, D. E. (undated). 'Inter-City Competition, the Community Growth Game', El Segundo, California: International Rectifier Paper.

Eilon, S. (1963). 'Management Games', *O. R. Quarterly*, vol. 14, no. 2, pp. 137–44.

Eldon, B. (1962). 'Business Games', in *Management Training Techniques*, British Institute of Management publication, pp. 23–32.

Entelek Incorp. (1964). 'Low Bidder Kit', developed by William R. Park, Newburyport, Massachusetts: Entelek Incorp.

Enzer, S., Gordon, T. J., Rochberg, R. and Buchela, R. (1969). *A Simulation Game for the Study of State Policies*, Report 9, Institute for the Future, Middletown, Connecticut.

Fairhead, J. (1965). 'The Validation of Business Exercises', *ATM Bulletin*, vol. 4.

Fairhead, J. N., Pugh, D. S. and Williams, W. J. (1965). *Exercises in Business Decisions: A Manual for Management Education*, London: E.U.P. Limited.

Featherstone, D. F. (1962). *War Games*, London: Stanley Paul.

Feldt, A. G. (1965). 'The Community Land Use Game', Ithaca, New York, Miscellaneous Papers no. 3, Division of Urban Studies, Center for Housing and Environmental Studies, Cornell University (mimeo).

Feldt, A. G. (1966*a*). 'Operational Gaming in Planning Education', *Journal of the American Institute of Planners*, January.

Feldt, A. G. (ed.) (1966*b*). 'Selected Papers on Operational Gaming', Miscellaneous Papers 5, Center for Housing and Environmental Studies, Division of Urban Studies, Cornell University, Ithaca, New York.

Feldt, A. G. (1966*c*). 'Potential Relationships between Economic Models and Heuristic Gaming Devices', a Paper prepared for presentation at a symposium on 'the role of economic models in policy formation' sponsored by the Office of Emergency Planning and the Department of Housing and Urban Development, 20 and 21 October.

Feldt, A. G. (1967*a*). 'Operational Gaming in Planning and Architecture', prepared for presentation at the A.I.A. Architects–Researchers Conference, Gatlinburg, Tennessee, 25 October (mimeo).

Feldt, A. G. (1967*b*). 'Gaming Techniques as a Communications Bridge between Systems Analysis and Public Administration', Paper prepared for presentation at the Annual Meeting of the American Association for the Advancement of Science: General Systems and Urban Planning, New York City, 30 December (mimeo).

Firey, W. (1947). *The Land Uses of Central Boston*, Cambridge, Massachusetts: Harvard University Press.

Fisher, H. B. (1967). 'The Renewal of Urban Land: Process, Decisions, Simulation', Chapel Hill, North Carolina: Center for Urban and Regional Studies, Institute for Research in Social Science, University of North Carolina.

Forbes, J. (1963). 'The College and University Planning Game', Academic Planning Tool Center, New Mexico State University, 1963.

Forbes, J. (1965). 'Operational Gaming and Decision Simulation', *Journal of Educational Measurement*, vol. 2, no. 1, June.

Friend, J. K. & Jessop, W. N. (1969). *Local Government and Strategic Choice: an operational approach to the processes of Public Planning*, London: Tavistock Publications.

Frye, F. F. (1970). 'Route Location Game', *Traffic Quarterly*, Saugatuck, Connecticut, January.

Gamson, W. A. (1966). *SIMSOC: A Manual for Participants*, Ann Arbor: Campus Publishers.

Gamson, W. A. (1969). *SIMSOC, Simulated Society: Instructor's Manual*, New York: The Free Press, Collier-Macmillan.

Garvey, D. M. (1965). *A Simulation of American Government*, Division of Social Sciences, Kansas State Teachers College, Emporia, Kansas, November.

Garvey, D. M. (1967). *Simulation, Role-Playing, and Sociodrama in Social Studies*, Emporia State Research Studies, Kansas State Teachers College, Emporia, Kansas, vol. XVI, no. 2, December.

Garvey, D. M. and Seiler, W. H. (1966). ' A Study of Effectiveness of Different Methods of Teaching International Relations to High School Students ', Final Report of Co-operative Research Project, no. 5–270, Emporia, Kansas (mimeo).

Geiger, M., Lawson, B. R., Schran, H. and Taylor, J. L. (1968). ' Some Recent European Developments of Instructional Simulation Systems in the Study of Urban Planning, *SCUPAD Bulletin* no. 5 Proceedings of the Third Salzburg Congress held at the Schloss Leopoldskron, 7–11 June, pp. 74–83.

Goldhammer, H. and Speier, H. (1959). ' Some Observations on Political Gaming ', *World Politics*, vol. XII, October, pp. 71–83.

Goldman, T. A. (1966). ' War Gaming, the War on Poverty ', Fort Washington, Pennsylvania: Philco Corporation, Unpublished Paper.

Good, I. J. (1954). ' Symposium on Monte Carlo Methods ' discussion, London: *Journal of the Royal Statistical Society*, vol. 16, pp. 68–9.

Goodman, R. F. (1968). ' The Inter-Community Simulation: An Experiment in Simulation-Gaming ', Los Angeles, California: Department of Political Science, University of California at Los Angeles (mimeo).

Gould, P. (1963). ' Man against his Environment: A Game Theoretic Framework ', Annals of the Association of American Geographers, pp. 290–7.

Gray, J. (1970). ' The Broadcasting of Game and Simulation Exercises ', in Armstrong, R. H. R. and Taylor, J. L. (eds.) *Instructional Simulation Systems in Higher Education*, pp. 120–6.

Greene, J. R. and Sisson, R. L. (1959). *Dynamic Management Decision Games*, New York: John Wiley and Sons, Inc.

Greenlaw, P. S., Herron, L. W. and Rawson, R. H. (1962). *Business Simulation in Industrial and University Education*, Englewood Cliffs, N.J.: Prentice Hall Inc.

Greenwald, H. A. (1966). ' The Scope and Limitation of Dynamic Games in Management Education ', Ph.D. Thesis submitted to the Victoria University, Manchester.

Griffin, S. F. (1965). *The Crisis Game: Simulating International Conflict*, Garden City, New York: Doubleday and Co.

Grundstein, N. D. (1961). ' Computer Simulation of a Community for Gaming ', Paper delivered at the annual meeting of the American Association for the Advancement of Science, Denver, Colorado, 29 December.

Grundstein, N. D. (1967). 'Some Conceptual Problems in the Simulation of Public Social Systems', in Feldt, A. G. (ed.), Selected Papers on Operational Gaming, Misc. Papers no. 5, Cornell University, Division of Urban Studies of the Center of Housing and Environmental Studies.

Grundstein, N. D. and Kehl, W. B. (1959). 'The Pittsburgh Community Model Game: A Proposal to Establish a Community Model Game with a Computer', University of Pittsburgh (mimeo).

Guetzkow, H. (ed.) (1962). *Simulation in Social Science*, Englewood Cliffs, N.J.: Prentice Hall Inc.

Guetzkow, H. (1964). *Simulation in International Relations*, Proceedings of the I.B.M. Scientific Computing Symposium on Simulation Models and Gaming, held on 7–9 December, at the Thomas J. Watson Research Center, Yorktown Heights, N.Y.

Guetzkow, H. *et al.* (1963). *Simulation in International Relations: Developments for Research and Teaching*, Englewood Cliffs, N.J.: Prentice Hall Inc.

Hale Committee (1964). *Report of the Committee on University Teaching Methods*, University Grants Committee, London: H.M.S.O.

Hall, P. (1967). 'New Techniques in Regional Planning: Experience of Transportation Studies', *Regional Studies*, vol. 1, pp. 17–21.

Hansen, W. B. (ed.) (1966). *Planning Research 1966*, see especially Hightower, H. C., 'A Role for Activity Analysis in Programming Public Expenditures', Washington D.C.: American Institute of Planners.

Harman, H. H. (1961). 'Simulation: A Survey', SP-260, System Development Corporation, Santa Monica, California, July.

Harris, B. (1965). 'Urban Development Models: New Tools for Planning', An introduction to a special issue on the above subject of the May edition of the *Journal of the American Institute of Planners*, vol. XXXI, pp. 90–4.

Hartman, J. J. (1966). 'Annotated Bibliography on Simulation in the Social Sciences', Rural Sociology Report no. 53, Iowa State University, Ames, Iowa.

Helmer, O. (1965). 'Social Technology', Rand Corp. Document no. P-3063, Santa Monica, California.

Hemphill, J. K., Griffiths, D. E. and Frederiksen, N. (1962). *Administrative Performance and Personality: A Study of the Principal in a Simulated Elementary School*, New York: Teachers College, Columbia University.

Henderson, R. (1968). 'Gaming as a Teaching Vehicle', A paper presented at the Heriot-Watt University and Edinburgh College of Art Department of Town and Country Planning Seminar on 'Teaching Methods in Scottish Planning Schools', 30 November.

Hendricks, F. H. (1960). 'Planning Operational Gaming Experiment', A paper presented to the North California Chapter of the A.I.P. Meeting on 'New Ideas in Planning', 19 November.

Hendricks, F. H. (1964). 'Micro Game Learning Model of San Francisco Housing Market' (unpublished paper).

Hendricks, F. H. *et al.* (1966). 'G.S.P.I.A. Management Planning Decision Exercise', Department of Urban Affairs Graduate School of Public and International Affairs, University of Pittsburgh, Pittsburgh, Pennsylvania, November.

Hoggatt, A. C. and Balderston, F. E. (1963). *Symposium on Simulation Models: Methodology and Applications to the Behavioral Sciences*', Cincinnati, Ohio: South Western Publishing Co.

Hoinville, G. (1970). 'Economic Evaluation of Community Priorities', Paper prepared for the 'Research for Social Policy Seminar', February.

Hoinville, G. and Barthould, R. (1969). 'Value of Time: Development Project Stage 2 Report', London: Social and Community Planning Research (mimeo).

Holland, E. P. (1965). 'Principles of Simulation', *American Behavioral Scientist*, vol. 9, September, pp. 6–10.

Hooper, R. (1968). 'Play the game – U.S. Style', *The Times Educational Supplement*, Friday, 9 August, p. 265.

House, P. (1968). 'Wedding Systemic and Role-Playing Models in Urban Research', Proceedings of the Seventh Symposium on Gaming held at the Asilomar Conference Grounds, Pacific Grove, California, 28–30 April.

House, P. and Patterson, P. D. (1969). 'An Environmental Gaming-Simulation Laboratory', *Journal of the American Institute of Planners*, vol. XXXV, no. 6, December, pp. 383–8.

Inbar, M. (1969). 'Towards a Sociology of Autotelic Behavior', Baltimore, Maryland: Johns Hopkins University Research Paper (mimeo).

Instructional Simulations Inc. (1969). 'I.S.I. Instructional Simulations Announcement', Newport, Minnesota.

Isard, W. (1967). 'Game Theory, Location Theory and Industrial Agglomeration', Papers of the Regional Science Association, vol. 18.

Isard, W. and Smith, T. E. (1967). 'Location Games: with Applications to Classic Location Problems', Papers of the Regional Science Association, vol. 19.

Jackson, J. R. (1960). 'Business Gaming in Management Science Education', in Churchman *et al.* (eds.), *Management Sciences; Models and Techniques*, vol. 1, London: Pergamon Press.

Kaplan, A. (1964). *The Conduct of Inquiry*, San Francisco: Chandler Publishing Co.

Kaplan, A. (1966*a*). ' Pollution and Neighborhood ', Cambridge, Massachusetts: Abt Associates Inc. (mimeo).

Kaplan, A. (1966*b*). ' The Game of Section ', Cambridge, Mass.: Abt Associates Inc. (mimeo).

Kelly, B. (1966). ' Introduction to Selected Papers on Operational Gaming ', Feldt, A. G. (ed.), Miscellaneous Papers no. 5, Center for Housing and Environmental Studies, Division of Urban Studies, Cornell University, Ithaca, New York.

Kibbee, J. M., Craft, C. J. and Nanus, B. (1961). *Management Games: New Technique for Executives*, Reinhold, New York.

Kilbridge, M. F. (1968). ' The Foundations of Urban Planning Models ', *Ekistics*, vol. 26, no. 155, October.

Kitchen, D. (1970). ' Games in Industrial and Management Training ', in Armstrong, R. H. R. and Taylor, J. L. (eds.), *Instructional Simulation Systems in Higher Education*, pp. 180–202.

Klietch, R. G. (1969). ' An Introduction to Learning Games and Instructional Simulation ', Newport, Minnesota: Instructional Simulations and Co.

Kraft, I. (1967). ' Opinions Differ: Pedagogical Futility in Fun and Games ', *N.E.A. Journal*, LVI, January, pp. 71–2.

Kriesis, P. (1968). ' Planning: A Rearguard View ', *Journal of the Town Planning Institute*, vol. 54, no. 5, May, pp. 226–7.

Laing's, J. Ltd. (undated). ' Exercise Quintain ', London: John Laing's Ltd (mimeo).

Lassiere, A. (1969). ' Transport and Amenity ', Paper prepared for the Seminar on the Treatment of Land, Infrastructure and Amenity in Cost Benefit Studies, Sunningdale, 10–11 September.

Laulicht, J. and Martin, J. (1966). ' The Vietnam War Game ', *New Society*, 27 January, pp. 6–8.

Lawson, B. (1968). ' Gaming Simulation in Planning ', Ljubljana, Yugoslavia: American–Yugoslav Project Seminar Paper S.5.1, February.

Lenel, R. M. (1969). *Games in the Primary School*, London: University of London Press Ltd.

Lewis, P. O. and Parry, R. A. (1966). ' The WESGAS Game – A Computer-Based Aid to Management Training ', British Joint Computer Conference, Eastbourne, 1966 (Organized by the Institute of Electrical Engineers).

Loewenstein, L. K. (1966). ' On the Nature of Analytical Models ', *Urban Studies*, vol. 3, no, 2, June, pp. 112–19.

Long, N. E. (1958). ' The Local Community and an Ecology of Games ', *American Journal of Sociology*, 64, November, pp. 251–61.

Loveluck, C. (undated). ' A Bibliography and Analysis of Business Games ', London: B.L.I.T.A.

Loveluck, C. (undated). 'How to Conduct a Business Game', London: B.L.I.T.A.

Luce, R. D. and Raiffa, F. (1957). *Games and Decisions*, New York: John Willey and Sons Inc.

McGlothlin, W. H. (1958). 'The Simulation Laboratory as a Developmental Tool', Santa Monica, California: Rand Corporation, p. 1454.

McKenney, J. L. (1962). 'An Evaluation of a Business Game in an M.B.A. Curriculum', *The Journal of Business*, xxxv, July, pp. 278–86.

McKenney, J. L. (1967). *Simulation Gaming for Management Development*, Boston: Division of Research, Harvard Graduate School of Business Administration.

McLeish, J. (1968). *The Lecture Method*, Cambridge Monographs on Teaching Methods no. 1, Cambridge: Institute of Education.

McLeish, J. (1970). 'Systems, Models, Simulations and Games in Education: A Description and Bibliography', in Armstrong, R. H. R. and Taylor, J. L. (eds.), *Instructional Simulation Systems in Higher Education*, pp. 9–20.

McLoughlin, J. B. (1966). 'The PAG Report: Background and Prospect', *Journal of the Town Planning Institute*, vol. 52, no. 7, July/August.

McLoughlin, J. B. (1967). 'A Systems Approach to Planning', Report of Proceedings, Town and Country Planning Summer School.

McLoughlin, J. B. (1968). 'Managing the Urban System', *Local Government Chronicle*, 26 October.

McLoughlin, J. B. (1969). *Urban and Regional Planning: A Systems Approach*, London: Faber.

MacNair, M. P. (ed.) (1954). *The Case Method at Harvard Business School*, Cambridge, Mass.: Harvard University Press.

Macunovich, D. (1967). 'Land Use/Transportation Simulation', in *Planning Games*, Proceedings of a P.T.R.C. Seminar held at the University of Manchester, 14 November, London: P.T.R.C., pp. 13–29.

Mahoney, T. A., Jerdee, T. H. and Korman, A. (1960). 'An Experimental Evaluation of Management Development', *Personal Psychology*, 13, pp. 81–9.

Malcolm, D. G. (1958). 'The Use of Simulation in Management Analysis – A Survey and Bibliography', S.P. 88, Santa Monica, California: System Development Corporation, November.

Malcolm, D. G. (1960). 'A Bibliography on the Use of Simulation in Management Analysis', *Operations Research*, VIII, March, pp. 169–77.

Mallen, G. L. (1970). 'Organisation Simulation – An Approach and A Method', in Armstrong, R. H. R. and Taylor, J. L. (eds.), *Instructional Simulation Systems in Higher Education*, pp. 170–9.

Mallen, G. L. (1970). 'Ecogame', Computer Arts Society, British Computer Society Specialist Group Announcement (mimeo).

Massey, D. (1968). *Problems of Location : Game Theory and Gaming-Simulation,* CES-WP-15, London: Centre for Environmental Studies, August.

Mayne, J. W. (1966). ' Glossary of Terms Used in Gaming and Simulation ', *Journal of the Canadian Operational Research Society,* vol. 4, July, pp. 114–118.

Mayo, E. (1957). *The Social Problems of an Industrial Civilization,* London: Routledge and Kegan Paul.

Meals, D. W. (1969). ' Games As Communication Devices ', in Proceedings of the National Gaming Council's Eighth Symposium, pp. 22–24, June.

Meehan, E. J. (1968). *Explanation in Social Sciences : A System Paradigm,* Homewood, Illinois: Dorsey Press.

Meier, R. L. (1961*a*). ' Explorations in the Realm of Organization Theory IV: The Simulation Organization ', *Behavioral Science,* VI, July, pp. 232–48.

Meier, R. L. (1961*b*). ' Teaching Through Participation in Micro-Simulation of Social Systems ', Paper delivered at the annual meeting of the American Association for the Advancement of Science, Denver, Colorado, 29 December.

Meier, R. L. (1963). ' " Game " Procedure in the Simulation of Cities ', in L. J. Duhl (ed.), *The Urban Condition,* New York: Basic Books Inc.

Meier, R. L. (1965). ' Simulations for Urban Planning ', Paper prepared for presentation at the Joint Planning Conference of the American Society of Planning Officials and the Community Planning Association of Canada, Toronto, Canada, 25–29 April.

Meier, R. L. (1968*a*). ' A Research Proposal: Simulations of Community and Social Development ', Berkeley, California: College of Environmental Design, Berkeley (mimeo).

Meier, R. L. (1968*b*). ' A General Systems Party ', *General Systems,* vol. XIII, pp. 209–12.

Meier, R. L., Blakelock, E. H. and Hinomoto, H. (1964). ' Simulation of Ecological Relationships ', *Behavioral Science,* vol. 9, no. 1, January, pp. 67–84.

Meier, R. L. and Duke, R. D. (1966). ' Gaming Simulation for Urban Planning ', *Journal of the American Institute of Planners,* vol. XXXII, no. 1, January, pp. 3–16.

Mitchell, C. R. (1971). ' The Use of Videotape and Close Circuit TV in Multi-Team Crisis Simulations ', in Armstrong, R. H. R. and Taylor, J. L. (eds.), *Feedback on Instructional Simulation Systems.*

Mitchell, N. B. (1968). ' Prospectus of Urban Planning Simulation – The Game ', Cambridge, Massachusetts: Neal B. Mitchell Assoc. (mimeo).

Mitchell, R. B. (1961). ' The New Frontier in Metropolitan Planning ', *Journal of the American Institute of Planners,* August.

Moise, E. (1964). 'The New Mathematics Program', in de Grazia, A. and Sohn, D. A. (eds.), *Revolution in Teaching*, New York: Bantam Books, pp. 171–87.

Monroe, M. W. (1968). 'Games as Teaching Tools: An Examination of the Community Land Use Game', Papers on Gaming Simulation no. 1, Center for Housing and Environmental Studies, Division of Urban Studies, Cornell University, Ithaca, New York.

Montessori, M. (1909). *The Montessori Method*, New York: Schocken.

Montessori, M. (1914). *Dr Montessori's Own Handbook*, New York: Schocken.

Montessori, M. (1965). *Spontaneous Activity in Education*, New York: Schocken.

Moreno, J. L. (1947). *The Theater of Spontaneity*, translated from the German, *Das Stagrufttheater*, New York: Beacon House.

Moss, J. M. (1958). 'Commentary on Harling's "Simulation Techniques in Operational Research"', *Operations Research*, July–August.

Murray, H. J. R. (1913). *A History of Chess*, Oxford, England: Oxford University Press.

Murray, H. J. R. (1952). *A History of Board-Games*, London: Oxford University Press, p. 267.

Nagelberg, M. (1970). *Simulation of Urban Systems – A Selected Bibliography*, WP-3, Institute for the Future, Middletown, Connecticut.

Nagelberg, M. (1970). *Selected Urban Simulations and Games*, Working Paper no. 4, Institute for the Future, Middletown, Connecticut.

National Gaming Council (1962 onwards). *Proceedings of the Symposium on Gaming*, held throughout the United States on an annual basis and the proceedings published by the host agency. (The Council was formed in 1962 as the East Coast War Games Council and adopted the current title in 1967.)

National Symposium on Management Games (Proceedings of) (1959). Center for Research in Business, University of Kansas, Lawrence, Kansas.

Naylor, T. H. (1969). 'Bibliography 19: Simulation and Gaming', *Computing Reviews*, vol. 10, no. 1, January, pp. 61–9.

Negrophonte, N. (1970). *The Architecture Machine: Towards a More Human Environment*, Cambridge, Mass.: M.I.T. Press.

Neumann, J. V. and Morgenstern, O. (1944). *Theory of Games and Economic Behavior*, Princeton, New Jersey: Princeton University Press.

Newbold, G. D. (1964). 'Business Games and the Theory of the Firm', Unpublished M.A. Thesis (University of Sheffield).

O'Connor, D. J. (1957). *An Introduction to the Philosophy of Education*, London: Routledge and Kegan Paul.

Ohm, R. E. (1966). 'Gamed Instructional Simulation: An Exploratory Model', *Educational Administration Quarterly*, April 1966, pp. 110–22.

Orlando, J. A. and Pennington, A. J. (1969). '" Build " – A Community Development Simulation Game', Philadelphia, Pennsylvania: Drexel Institute (mimeo).

Osgood, C. E. *et al.* (1965). *The Management of Meaning*, Urbana: University of Illinois Press.

Paterson, T. T. (1970). 'The Use of Simulation Exercises in the Teaching of Industrial Relations', in Armstrong, R. H. R. and Taylor, J. L. (eds.), *Instructional Simulation Systems in Higher Education*, pp. 112–19.

Patterson, J. R. P. (1968). 'Gaming Sessions', Summary of a workshop session held at the A.I.P.'s 51st Annual Conference in Pittsburgh, 12–15 October, A.I.P. Newsletter, November, vol. 3, no. 11, pp. 3–4.

P-E Consulting Group Ltd (1966). 'Hospital Exercise in Long Term Planing' (A Simulation Exercise prepared for the Training Department of Wessex Regional Hospital Board), January.

Pennington, A. J. (1969). 'Uses and Misuses of Computers in Urban Affairs' (A paper prepared for the 1969 Joint Computer Conference, 14–16 May, Boston, Mass.) (mimeo).

Pestalozzi, J. H. (1819 onwards). There are three comprehensive editions of Pestalozzi's work, the latter edition is still not available in its entirety:

1. *Pestalozzis sämtliche Schriften*, I. G. Lottasche Buchhandlung, Stuggart and Tübengen, 1819–26.
2. *Pestalozzis sämtliche Werke* (2nd edition), edited by L. W. Seffarth, Liebnitz, 1899–1902.
3. *Pestalozzis sämtliche Werke*, edited by A. Buchenau, E. Spranger, and H. Strettbacher, Berlin, Leipzig, Zurich, 1927.

A brief summary of these works is provided by Heafford, M. (1967) *Pestalozzi*, London: Methuen.

Peston, M. and Coddington, A. (1967). *The Elementary Ideas of Game Theory*, C.A.S. Occasional Paper no. 6, London: H.M.S.O.

Piaget, J. (1962). *Play, Dreams, and Imitation in Childhood*, New York, N.Y.: The Norton Library.

Piaget, J. (1965). *The Moral Judgement of the Child*. New York, N.Y.: The Free Press.

Pigors, P. and Pigors, F. (1961). *Case Method in Human Relations: the Incident Process*, New York: McGraw Hill.

Piper, P. J. W. and Rae, J. (1969). 'Madingley Game 1969', A paper prepared for the Conference on 'The Diploma in Art and Design' held at Madingley Hall, Cambridge, March.

Plattner, J. W. and Herrow, L. W. (1962). *Simulation: Its Use in Employee Selection and Training*, A.M.A. Management Bulletin, no. 20, New York: American Management Association.

Polycon (1967). 'Proposal for the Use of the Factoplan Training Seminar', London: Polycon Group Building Industry Consultants (mimeo).

Press, C. and Adrian, C. R. (1966). 'Choice Theory: A Glossary of Terms', Technical Bulletin B-55, Lansing, Michigan: Institute for Community Development, Continuing Education Service, Michigan State University, June.

Project Simile (1965 onwards). *Occasional Newsletter About Uses of Simulations and Games for Education and Training*, La Jolla, California: Western Behavioral Sciences Institute (mimeo).

Project Simile (1966). 'An Inventory of Hunches about Simulations as Educational Tools', La Jolla, California: Western Behavioral Sciences Institute, May.

P.T.R.C. (Planning and Transport Research and Computation Co. Ltd) (1967). '*Planning Games*', Proceedings of a P.T.R.C. Seminar held at the University of Manchester, 14 November, London: P.T.R.C.

P.T.R.C. (Planning and Transport Research and Computation Co. Ltd) (1968). 'Land Use/Transportation Simulation Kit', London: P.T.R.C.

Pugh, D. (1965). 'Games and Exercises – A Comment on Terminology', in Life, A. and Pugh, D. (eds.), *Business Exercises: Some Developments*, A.J.M. Occasional Paper no. 1, February, Oxford: Blackwell.

Quade, E. S. (1964). 'Methods and Procedures', in Quade, E. S. (ed.), *Analysis for Military Decisions*, Chicago: Rand McNally and Co.

Rackham, N. (1970). 'The Effectiveness of Gaming and Simulation Techniques', in Armstrong, R. H. R. and Taylor, J. L. (eds.), *Instructional Simulation Systems in Higher Education*, pp. 203–10.

Rae, J. (1969). 'Games', *Architects Journal*, no. 15, vol. 149, 9 April, pp. 977–83.

Rae, J. (1970). 'Gaming, Decision Making, and Social Skills' in Armstrong, R. H. R. and Taylor, J. L. (eds.), *Instructional Simulation Systems in Higher Education*, pp. 141–69.

Raia, A. P. (1966). 'A Study of the Educational Value of Management Games', *Journal of Business* (University of Chicago), vol. XXXIX, no. 3, July, pp. 339–52.

Rapoport, A. (1960). *Fights, Games and Debates*, Ann Arbor: University of Michigan Press.

Rapoport, A. and Chammah, A. M. (1965). *Prisoner's Dilemma*, Ann Arbor: University of Michigan Press.

Raser, J. R. (1969). *Simulation and Society: An Exploration of Scientific Gaming*, Boston, Mass.: Allyn and Bacon.

Rauner, R. M. and Steger, W. A. (1961). 'Game-Simulation and Long-Range Planning', P-2355, Rand Corporation, Santa Monica, California, 22 June.

Ray, P. H., Duke, R. D. and Feldt, A. G. (1966). 'Gaming-Simulations for Transmitting Concepts of Urban Development', Prepared for presentation at the summer meetings of the American Sociological Association (mimeo).

Ray, P. H. and Duke, R. D. (1967). 'The Environment of Decision-Makers in Urban Gaming-Simulations', Paper delivered at the Symposium on 'Simulation Models of the Decision-Makers' Environment' held at Wayne State University, 12 May.

Redgrave, M. J. (1962). 'Some Approaches to Simulation, Modeling and Gaming at S.D.C.', Santa Monica, California: System Development Corporation, SP-721 (mimeo).

Reed, L. R. (1966). 'A Study of the Feasibility of Using Operational Simulation Techniques for Evaluating Administrative Skills Possessed by Instructional Communications Specialists', Syracuse, New York: Syracuse University.

Reiner, T. A. (1963). *The Place of the Ideal Community in Urban Planning*, Philadelphia: The University of Pennsylvania Press.

Reske, J. D., Gilon, P., Strong, A., Miller, L. and Miller J. (1966). 'Comexopolis Mark I Hand Book', Comex Research Project, University of Southern California, Los Angeles (mimeo).

Ricciardi, F. M. *et al.* (1957). *Top Management Decision Simulation: The A.M.A. Approach*, New York: American Management Association.

Riley, V. and Young, J. P. (1957). 'Bibliography on War Gaming', Operations Research Office, Johns Hopkins University, 1 April.

Rivett, B. H. P. (1961). 'Business Games as an aid to Management Strategy', *New Scientist*, no. 246, 3 August, pp. 264–6.

Robinson, J. A. (1966). 'Simulation and Games', in Rossi, P. H. and Biddle, B. J. (eds.), *The New Media and Education*, Chicago, Illinois: Aldine Publishing Co., pp. 85–123.

Rockwell *et al.* (1968). *Northeastern Illinois Planning Commission. The Plan Study: Methodology*, Phase two: the Comprehensive Plan, Chicago: N.I.P.C.

Roethlisberger, F. J. and Dickson, W. J. (1939). *Management and the Worker*, Cambridge, Mass.: Harvard University Press.

Rohn, W. E. (1964). *Führungsentscheidungen im Unternehmensplanspiel*, Essen: Kerlag W. Girardet.

Rosenhead, J. V. (1968). 'Experimental Simulation of a Social System', *Operational Research Quarterly*, September.

Rossi, P. H. (1957). 'Community Decision Making', *Administrative Science Quarterly*, vol. 1, March.

Schild, E. O. (1968). 'Interaction in Games' in Boocock, S. S. and Schild, E. O. (eds.), '*Simulation Games in Learning*', Beverly Hills, California: Sage Publications Inc.

Schran, H. (1964). 'Das Planspiel im Städtebau – ein Anwendungsbeispiel', *Stadtbauwelt*, Heft 3, pp. 174–9.

Schran, H. (1968). 'Planspiel – Simulation', *Stadtbauwelt*, Heft 19, pp. 1443–7.

Scott, A., Lucas, W. A. and Lucas, T. M. (1967). *Simulation and National Development*, New York: John Wiley and Sons.

Shaftez, F. R. and Shaftez, G. (1967). *Role-Playing for Social Values*, Englewood Cliffs, N.J.: Prentice Hall Inc.

Shephard, R. W. (1970). 'The Possibilities of Using War Games to Train Army Commanders', in Armstrong, R. H. R. and Taylor, J. L. (eds.), *Instructional Simulation Systems in Higher Education*, pp. 21–7.

Shubik, M. (ed.) (1954). *Readings in Game Theory and Political Behavior*, Garden City, New York: Doubleday Short Studies in Political Science.

Shubik, M. (1960). 'Bibliography on Simulation, Gaming, Artificial Intelligence and Allied Topics', *Journal of American Statistical Association*, December, vol. 55, pp. 736–51.

Shubik, M. (ed.) (1964). *Game Theory and Related Approaches to Social Behavior*, New York: John Wiley and Sons Inc.

Shubik, M. (1968). 'Gaming: Costs and Facilities', *Management Science*, vol. 14, no. 11, July, pp. 629–60.

Skeffington Committee (1969). *People and Planning*, London: H.M.S.O.

Skinner, B. F. (1953). *The Technology of Teaching*, New York: Appleton-Century Crofts.

Skinner, B. F. (1963). *Science and Human Behavior*, New York: Macmillan and Co.

Smith, G. A. and Cole, J. P. (1967). 'Geographical Games', *Bulletin of Quantitative Data for Geographers* no. 7, Department of Geography, University of Nottingham.

Smith, N. M. and Marney, M. (1961). 'Simulation and Gaming as Cognitive Models', Bethesda, Maryland: Operations Research Office of Johns Hopkins University Paper.

Specht, R. D. (1957). 'War Games', Santa Monica, California: Rand Corporation, 18 March, P-1041.

Sprague, H. T. and Shirts, G. R. (1966). 'Exploring Classroom Uses of Simulation', La Jolla, California: Western Behavioral Sciences Institute (mimeo).

Stavros, D. C. (1967). 'Planning Application of a Gaming-Simulation of Regional Development', Master of Regional and Community Planning Thesis, Manhattan, Kansas: Kansas State University.

Stea, D. (1967). 'Mediating the Medium', *Journal of the American Institute of Architects*, December, pp. 67–70.

Steinitz, C. and Rogers, P. (1968). *A Systems Analysis Model of Urbanization and Change: An Experiment in Interdisciplinary Education*, Cambridge, Mass.: Harvard Graduate School of Design.

Stevens, B. H. (1961). 'An Application of Game Theory to a Problem in Location Strategy', Papers and proceedings of the Regional Science Association, vol. 7.

Stones, E. (1968). *Learning and Teaching: A Programmed Introduction*, London: John Wiley and Sons.

Suiits, B. (1967). 'Is Life a Game We are Playing', *Ethics*, 77, pp. 209–13.

System Development Corporation (1965). 'Simulation: Managing the Unmanageable', Santa Monica: System Development Corporation Magazine, vol. 8, no. 4, April.

Systems Gaming Associates (1969). 'CLUG Information Service', vol. 1, Ithaca, New York: Systems Gaming Associates.

Tansey, P. J. and Unwin, D. (eds.) (1967 onwards). 'Bulletins on Academic Gaming and Simulation', Reading, Berkshire: Bulmershe College of Education (mimeo).

Tansey, P. J. and Unwin, D. (1969). *Simulation and Gaming in Education*, London: Methuen.

Taylor, J. L. (1967). 'A Synoptic View of Urban Phenomena', *Journal of Town Planning Institute*, vol. 53, no. 1, January.

Taylor, J. L. (1968*a*). 'Some Pedagogical Aspects of a Land Use Planning Approach to Urban System Simulation', Seminar Paper Series, Series A, no. 14, Geography Department, University of Bristol, p. 19.

Taylor, J. L. (1968*b*). 'Planning Games – Preliminary Observations on Some Teaching Applications in Developing Countries', Report of the Proceedings of the Town and Country Planning Summer School held at the University of Manchester, 4–15 September, pp. 84–6.

Taylor, J. L. (1969*a*). 'Some Aspects of an Instructional Simulation Approach to the Urban Development Process', Doctoral Dissertation, University of Sheffield.

Taylor, J. L. (1969*b*). 'Forecasting Futures', *Architectural Design*, vol. XXXIX, no. 10, pp. 568–9.

Taylor, J. L. (ed.) (1969*c*). Social Science Instructional Simulation Systems: A Selected Bibliography, SURISS Project Paper no. 4, Department of Town and Regional Planning, University of Sheffield.

Taylor, J. L. and Carter, K. R. (1967). 'Instructional Simulation of Urban Development: A Preliminary Report', *Journal of the Town Planning Institute*, vol. 53, no. 10, pp. 443–7.

Taylor, J. L. and Carter, K. R. (1969*a*). 'A Decade of Instructional Simulation Research in Urban and Regional Studies' in Armstrong, R. H. R. and Taylor, J. L. (eds.), *Instructional Simulation Systems in Higher Education*, pp. 45–66.

Taylor, J. L. and Carter, K. R. (1969*b*). 'Some Instructional Dimensions of Urban Gaming-Simulation', University of Newcastle-on-Tyne, Department of Geography Seminar Papers no. 7, March.

Taylor, J. L. and Carter, K. R. (1971). 'Some Assessments of Gaming-Simulation Systems Used in Studying the Urban Development Process', in Armstrong, R. H. R. and Taylor, J. L. (eds.), *Feedback on Instructional Simulation Systems*, Cambridge: Cambridge Monographs on Education no. 5.

Taylor, J. L. and Geiger, M. (1967). 'Ein Laboratorium für den Planner', *Werk*, Zurich, no. 9, pp. 581–3.

Taylor, J. L. and Geiger, M. (1968). 'Erste Erfahrungen mit Spiel-Simulationen im Planerunterricht', *Werk*, Zurich, vol. 53, no. 10, October, pp. 686–9.

Taylor, J. L. and Maddison, R. N. (1968). 'A Land Use Gaming Simulation: The Design of a Model for the Study of Urban Phenomena', Beverly Hills, California: *Urban Affairs Quarterly*, vol. 3, no. 4, June, pp. 37–51.

Teaching Reference Community Unit (1966). 'Dixon, Tiller County, U.S.A.' (Teaching Reference Community), Atlanta, Georgia, U.S. Department of Health, Education and Welfare.

Teaching Research *et al.* (1967). 'Instructional Uses of Simulation: A Selected Bibliography', Prepared by Teaching Research, a division of the Oregon State System of Higher Education, Monmouth, Oregon, in co-operation with North West Regional Educational Laboratory, Portland, Oregon.

Teaching Research *et al.* (1967). 'Of Men and Machines: Supplementary Guide', Teaching Research, a division of the Oregon State System of Higher Education, Monmouth, Oregon, in co-operation with North West Regional Educational Laboratory, Portland, Oregon.

Thomas, C. J. and Deemer, W. L. (1957). 'The Role of Operational Gaming in O.R.', *Operations Research*, vol. 5, no. 1, February, pp. 1–27.

Thomas, C. J. and McNichols, G. R. (1969). 'Gaming: 1959–1979 (Why People Play Games – Report of a Survey)', in Proceedings of the National Gaming Council's Eighth Symposium, 22–24 June.

Thorelli, H. B. and Graves, R. L. (1964). *International Operations Simulation*, London: Collier Macmillan.

Town Planning Institute (1969). *Examinations Handbook (revised 1969)*, London: Town Planning Institute.

Traffic Research Corporation (1963). *Review of Existing Land Use Forecasting Techniques*, Boston, Mass.: T.R.C.

Traffic Research Corporation (1969). ' Merseyside Area Land Use and Transportation Study – Miscellaneous Reports including a Review of the Application of Gaming Techniques to Planning ', M.A.L.T.S. Technical Report no. 24, Liverpool: T.R. Corporation.

Travis, A. S. (1969), ' Ends and Means: Planning for a Changing Society ', An inaugural lecture given at Herriot Watt University, Edinburgh, on 24 April.

Turner, A. J. D. (1949). *Valentine's Sand Table Exercises*, Tenth Revised and Enlarged Edition, Aldershot: Gale and Polden Ltd.

Twelker, P. A. (1969). ' Instructional Simulation Systems: An Annotated Bibliography ', Corvallis, Oregon: Continuing Educational Publications.

Urban Systems Simulations (1968*a*). ' Region Manual ', Urban Systems Simulations at the Washington Center for Metropolitan Studies, p. 89.

Urban Systems Simulations (1968*b*). ' City I Player's Manual ', Urban Systems Simulations at the Washington Center for Metropolitan Studies, p. 26.

Van der Heijden, H. H. J. M. (1968). ' Een Simulatiespel ten behoeve van stedelijke planning ', in *Kwartaalbericht*, Eindhoven, 19e Jaargang, pp. XII–XV.

Walford, R. (1968). ' Six Classroom Games for Use in Geography Teaching ', London: Maria Grey College (mimeo).

Walford, R. (1969). ' Operational Games and Geography Teaching ', *Geography*, no. 242, vol. 54, part 1, January, pp. 34–42.

Walford, R. (1969). *Games in Geography*, London: Longmans.

Webb, P. C. and Wheeler, G. E. (1962). ' Operation Taurus: A Business Game Designed for the Building Industry ', A Record of the Proceedings at a Management Conference of Messrs. Howard Farrow Limited in September, *The Journal of Industrial Economics*, vol. x, no. 2, March.

Weiner, M. G. (1959). ' An Introduction to War Games ', Santa Monica, California: Rand Corporation.

Wiener, R. S. P. (1971). ' Simulation with the Use of Video-tape Equipment as a Research Tool ', in Armstrong, R. H. R. and Taylor, J. L. (eds.), *Feedback on Instructional Simulation Systems.*

Weiss, S. F., Smith, J. E., Kaiser, E. J. and Kenney, K. (1966). ' Residential Developer Decisions ', Chapel Hill, North Carolina: Center for Urban and Regional Studies, Institute for Research in Social Sciences, University of North Carolina.

Werner, R. and Werner, J. T. (1969). 'Bibliography of Simulations: Social Systems and Education', La Jolla, California: Western Behavioral Sciences Institute (mimeo).

Western Behavioral Sciences Institute (1969). 'Sitte', La Jolla, California: Western Behavioral Sciences Institute (mimeo).

Williams, J. D. (1954). *The Complete Strategyst*, New York: McGraw Hill.

Wilson, A. G. (1968). 'Models in Urban Planning: A Synoptic Review of Recent Literature', *Urban Studies*, vol. 5, no. 3, pp. 249–76.

Wilson, A. (1968). *The Bomb and the Computer*, London: Barrie and Rockliff, the Cresset Press.

Wing, R. L. (1968). 'Two Computer-Based Economics Games for Sixth Graders', in Boocock and Schild (eds.), *Simulation Games in Learning*, Beverly Hills, California: Sage Publications Inc., pp. 155–65.

Wolff, P. (1966). 'The Game of Empire', Occasional Paper no. 9, The Social Studies Curriculum Program, Educational Services Inc., Cambridge, Mass.

Wolpert, J. (1964). 'The Decision Process in Spatial Context', *Annals of the Association of American Geographers*, vol. 54, December, pp. 537–58.

Young, J. P. (1956). 'A Brief History of War Gaming', Washington D.C.: Operations Research Office, Johns Hopkins University.

Young, J. P. (1957). 'A History and Bibliography of War Gaming', Staff Paper ORO-SP-13, Washington D.C.: Operations Research Office, Johns Hopkins University.

Young, J. P. (1959). 'A Survey of Historical Developments in War Games', Staff Paper ORO-SP-98, Washington D.C.: Operations Research Office, Johns Hopkins University.

Zelditch, M. and Evan, W. (1962). 'Simulated Bureaucracies: A Methodological Analysis', in Guetzkow, H. (ed.), *Simulation in Social Science: Readings*, Englewood Cliffs, N.J.: Prentice Hall Inc.

Zieler, R. (1969). 'Games for School Use', Center for Educational Services and Research, Board of Co-operative Educational Services, Westchester, New York.

Zoll, A. A. (1966). *Dynamic Management Education*, Seattle, Washington: Management Education Associates.

Index